The Human Element

Issues about organizations, how they work, whether we can afford them, and whether they will survive at all, is front page news every day. The book puts the case that the human element is absolutely critical, and the reason why we have been losing faith in the effectiveness of organizations over the past generation, is that it has been deliberately excised. It has been removed because of human fallibility.

However, there is a growing understanding, not that people are infallible, nor that they are endlessly trustworthy and benevolent, but they are what makes change possible. *The Human Element* uses this idea to set out the ten New Rules for organizations, revealing where they are working already including the latest developments in ideas such as system thinking and co-production.

It explains the future in terms of the People Principle: "If you employ imaginative and effective people, especially on the frontline, and give them the freedom to innovate, they will succeed. If you don't, they will fail."

The Human Element is an essential read for managers in the public, private and voluntary sectors as well as the general public.

David Boyle is a Fellow of the New Economics Foundation and author of numerous books, including (as co-author) *The New Economics* (Earthscan 2009).

The Human Element

Ten new rules to kick-start our failing organizations

David Boyle

publishing for a sustainable future

London & New York

First published 2012
by Earthscan
2 Park Square, Milton Park, Abingdon, Oxon OX14 4RN

Simultaneously published in the USA and Canada
by Earthscan
711 Third Avenue, New York, NY 10017

Earthscan is an imprint of the Taylor & Francis Group, an informa business

© 2012 David Boyle

British Library Cataloguing in Publication Data
A catalogue record for this book is available from the British Library

Library of Congress Cataloging in Publication Data
Boyle, David, 1958-
The human element : ten new rules to kickstart our failing organizations /
David Boyle.
p. cm.
ISBN 978-1-84971-449-5 (hbk) -- ISBN 978-0-203-12826-8 (ebk)
1. Organizational change. 2. Communication in organizations.
3. Industrial management. I. Title.
HD58.8.B695 2011
658.4'06--dc23
2011033510

ISBN: 978-1-84971-449-5 (hbk)
ISBN: 978-0-203-12826-8 (ebk)

Typeset by Saxon Graphics Ltd, Derby

Printed and bound in Great Britain by the MPG Books Group

To Edgar and Chris Cahn
with thanks, admiration and love

The human voice is the most beautiful instrument of all, but it is the most difficult to play.

(Richard Strauss)

Table of contents

Acknowledgements

One of the main reasons I ever started thinking about this book was the tiny lavatory in the busiest railway interchange in Europe, Clapham Junction, a station I seem to have measured out my life in. In the summer of 2004, South West Trains (SWT) replaced the loo with three booths through the car park, down the road and round the corner, one of which required a disabled key to open.

It set me thinking about those human requirements that seem to be in the process of being ironed out in the name of efficiency. If SWT's customers had been one-dimensional machines, two booths somewhere else might have been adequate. As it was, people tended to use the car park, if they dared. Perhaps that's what was always intended.

I don't buy the idea behind the slogan 'broken Britain'. Britain is not broken, as a walk down the vast majority of streets will tell you, and this book is an example of how it works. It was written with the help of a whole range of interlocking networks of people that are the very opposite of broken. This is a way of saying I am enormously indebted to huge numbers of people, many of whom probably won't be aware of it. They start with Edgar and Chris Cahn at the Time Dollar Institute in Washington, Martin Simon and Martin Farrell at Time Banking UK, and all those networks of time bankers and co-production pioneers on both sides of the Atlantic who gave rise to so much of it.

The second group of people who have been central are my colleagues at the New Economics Foundation, especially all those who have worked in the co-production team, Sherry Clark, Karina Krogh, Karen Lyon, Claire Navaie, Ben Robinson, Karen Smith and especially Lucie Stephens and Julia Slay. But also those in other teams, who have been hugely influential, from Ed Mayo who originally brought me in to work there, Stewart Wallis who still lets me work there, Anna Coote for her thinking and advice, Ruth Potts for all her help with my localism thinking, Josh Ryan-Collins for his work on commissioning and Andrew Simms and Lindsay Mackie for their friendship, advice and encouragement, and many others.

The third group is mainly friends in politics, especially Nick Clegg, whose conversation was a key influence when I was first planning the book, but also Richard Kemp and his team at the Local Government Association. These are Liberal Democrats, but the work of Zoe Gannon and Neal Lawson at Compass have been enormously influential, as have Phillip Blond (who I've met) and Jill Kirby of the Centre for Policy Studies (who I haven't).

The fourth group is people I have encountered around the Town and Country Planning Association, including Tony Gibson and Colin Ward, the pioneers of UK community development, and especially Nick Matthews for letting me write my 'Localism' column in *Town & Country Planning* where many of these ideas first appeared.

Other people who have provided invaluable advice or time include Titus Alexander, John Ashcroft, Becky Booth, Pat Brown, Duncan Brack, Louise Casey, Barry Cooper, Corrina Cordon, Anna Crispe, Mike Dixon, Keri Facer, Michael Fielding, Olly Grender, Jeremy Hargreaves, Judith Hodge, Amanda Horton-Mastin, Victoria Johnson, Naresh Khatri, Abby Letcher, Polly Mackenzie, Becky Malby, Debbie Morrison, Julia Neuberger, Sophia Parker, Kathy Perlow, Diane Plamping, John Rafferty, Zoe Reed, Samir Rihani, Bernie Rochford, John Seddon, Chris Seeley, John Sharkey, Marian Storkey and Polly Wiessner, and many others, some of whom preferred not to be named. I also want to thank Mats Lederhausen, from the consultancy BeCause, who inspired me to embark on this venture and introduced me to the work of Dov Seidman, which has also been hugely influential.

My two boys, Robin and William, have played a key part, putting up with me working in my hut above their sandpit, but also teaching me more about being human, which was important for writing this book. It is conventional also to thank your partner for their patience at this point. But Sarah has been more than merely patient. There is hardly a thought in this book which she hasn't inspired, either by her own work in public services – or by her pioneering work on co-production – but also because of her unique take on the world, from which I learn a little more every day. This is as much her book as it is mine.

Introduction:
The People Principle

Two things are infinite: the universe and human stupidity; and I'm not sure about the universe.

(Albert Einstein)

Don't automate, obliterate!

(Michael Hammer, *Harvard Business Review*, 1990)

Summary

- The human element may be a source of error, but it is also the only source of genuine change.
- If you employ imaginative and effective people, especially on the frontline, and give them the freedom to innovate, they will succeed. If you don't, they will fail.

It is the evening of 1 August 1798, in a sticky Mediterranean dusk, and Horatio Nelson's Mediterranean Fleet has finally tracked down their French opponents at anchor in Aboukir Bay on the Egyptian coast. He is determined to bring them to action, even in the gathering gloom. The British gun crews are crouching by their cannon while their French counterparts heave their heavy armaments onto the seaward side where the British will come. Ahead of Nelson is a more powerful fleet, commanded by François-Paul Bruey in his flagship *L'Orient*, bigger than any ship in the Royal Navy at the time. The Battle of the Nile is about to begin.

Nelson had prepared for this battle by setting out clear rules of engagement, discussed with his captains evening after evening around his table on the *Vanguard*. They would attack Bruey's front and centre and overwhelm them before the rear could take a decisive part in the battle. That was the broad plan; the details would have to take care of themselves as circumstances arose, and he trusted his captains to interpret the plan effectively.

Nelson was no disciplinarian, and he had already gained a reputation for disobeying orders during the Battle of Cape St Vincent. He had steered out of line because he saw the chance to cut off a group of Spanish ships from the rest and managed to capture them. Even if this wasn't explicit, his captains knew this was his style and it was what he expected of them – not slavish obedience to detail, but enthusiastic commitment to the objective. These regular dinners were the beginning of the trusting collegiate atmosphere he managed to instil among his commanders, which gave rise to the idea of a 'band of brothers'. It was a quotation from the speech that Shakespeare put into the mouth of Henry V before Agincourt, and intended to convey a similar burst of military inspiration.

Even so, there was naturally some rivalry within this band, and since Captain Thomas Foley in the *Goliath* happened to be leading the line when the French came into sight, he ordered his men to get the battle sails ready, so that he could stay in front when the order came to get into line of battle. So it was Foley, standing next to his helmsman, the battle ensigns flying behind him, who saw the emerging opportunity as the disposition of the French ships became clear. There they were anchored along the shore, and he realized there might just be enough space to squeeze along their undefended side, between the French line and the shore itself.

It was a risky decision. Thinking fast as the battle got ever nearer, Foley realized that Bruey's ships must have anchored with enough space to swing round at anchor as the tide changed, so there would almost certainly be enough sea to avoid running aground. He also happened to have an old French atlas with a map of Aboukir Bay, showing what depth the water was. But there was no time to consult anyone else. Foley steered between the French ships and the shore leading the British line after him.

It was dangerous and potentially disastrous, but it was right. Foley's leap of faith was the decisive moment of the battle, which reached its crescendo in the middle of the night when a huge explosion tore apart the French flagship and put an end to Napoleon's ambitions in Egypt. Foley was rightly hailed as the hero of the victory at Aboukir Bay of which Nelson had been the architect.

So although Nelson laid down the framework for the battle, with regular dinners for his captains, making sure his intentions became second nature to them, Foley knew he was allowed to do something entirely different if he saw an opportunity. He was able to break with conventional thinking, and the apparent drift of his orders, and use his intuition. Would he have managed to win the battle if he had been governed by the management culture from British public services two centuries later? Hard to know, but probably not.

The point was that he knew he had to take the decision, knew he was expected to, felt confident to do so, and did so in style.

It wasn't that the Royal Navy at the time was somehow not hierarchical, or was a tolerant organization of equality and gentleness. It was as complex a management system as we have ever devised, but with these two exceptions. Once they were at sea, captains were largely on their own in terms of command. There was no chance of micromanagement from the Admiralty. The other factor was Nelson himself, who was able to cut through the hierarchy to build a real relationship with his captains. Foley was able to use his judgement brilliantly because his commander expected it of him. If he had failed to do so, the battle would have been less successful.

Fast-forward nearly a century to 22 June 1893, but again to the British Mediterranean Fleet, by then the decisive force in global military affairs. By that time, the Royal Navy revered the name of Nelson and paid lip service to his cult of structured disobedience – the telescope to the blind eye and everything that went with it – but had rather forgotten what it meant.

Nelson's successor as commander was Admiral Sir George Tryon, charming on the dinner party circuit but known as a dictatorial martinet when he was at sea. He tried to keep his intentions hidden from his subordinates to help them practice in unpredictable situations but, the day before this fatal incident, he had actually told his captains what he wanted to do. He was going to turn his two columns of ironclads towards each other before they anchored for the night. It was a risky manoeuvre. Some brave captains suggested that, given the turning circles of the ships, the columns ought to be at least 1600 yards apart when they started to turn. It wasn't quite clear whether Tryon had agreed.

When the time came, off the coast of what is now Lebanon, Tryon unexpectedly ordered the two columns to start turning when they were only 1000 yards apart. Two officers queried the order, but he snapped at them to get on with it. Admiral Hastings Markham, leading the other column, was confused by the dangerous signal and delayed his acknowledgement. 'What are you waiting for?' signalled Tryon.

What was going to happen seemed horribly apparent to everyone except Tryon, but nobody acted to prevent it. Three times, the flag captain of his flagship *Victoria* asked for permission to go astern as the two leading ships hurtled towards each other, but did nothing. Only at the last minute, as Markham's flagship *Camperdown* hurtled towards the *Victoria* with its ram below the waterline, Tryon shouted, 'Go astern, go astern!'

It was too late. There was a grinding crash as the *Camperdown*'s ram buried itself in the flagship. *Victoria* capsized and sank 13 minutes later. As many as 358 sailors lost their lives. One of them was Admiral Tryon, who was said to have appeared mysteriously to his wife and guests at a dinner party in Eaton Square at his moment of departure.

These two stories, both about the commanders of the British Mediterranean Fleet, provide a blueprint for different organizational styles. Companies or hospitals are not quite the same as fleets but, even so, organizations run by Nelsons tend to work, and those run by Tryons tend not to. Tryon organizations can get by, as we shall see, but never quite in the brilliant ways that the Nelson organizations do. Ironically, the ill-fated Tryon was always known as a brilliant and innovative strategist. His fatal flaw was his authoritarian style of leadership, which left those who would have been in his band of brothers – if he had been Nelson – fatally in the dark. They could see the details of what they were supposed to do, but not the big picture. There they were, flailing around, not daring to act to avoid disaster even though they could see it coming.

The situation was extreme, one ironclad with a ram bearing down on another, but it is also familiar. We have all worked for organizations where a similar culture prevails, where disaster looms and most people decide it is probably best to say nothing. We *could* say things that would improve services or performance or avoid accidents or disasters, but it is less risky to keep quiet. Maybe this doesn't matter so much running fast food franchises or corner shops, but in hospitals it matters very much indeed, and in the armed services it has huge implications.

These dilemmas are always more extreme when it involves the military – the avoidable disease during the Crimea, the appalling loss of life around Stalingrad, because nobody challenged the two vicious dictators directing the war. Not to mention the Somme or Vietnam or Iraq. When President Lyndon Johnson barricaded himself into the Pentagon in the mid-1960s, surrounded by military advisors, choosing targets for his B52s from 12,000 miles away, he was making a familiar mistake.

After the disastrous Cambrai offensive in 1917, it emerged that the British commander's intelligence officers knew that the Germans had been reinforced from the Russian Front, but deliberately kept him in the dark. They did so, said Brigadier Charteris, because they 'did not wish to weaken the commander-in-chief's resolution to carry on with the attack'.

You might recognize a similar pattern to these among those who presided over Stafford Hospital and the avoidable 400 deaths there or the people who overloaded the banking system with worthless sub-prime debt.

The psychologist Norman Dixon spent the 1970s carrying out a fascinating study of the worst commanders in history, published as *On the Psychology of Military Incompetence*, and found that they had a great deal in common. They were often repressed personalities, who promoted sycophants and hugely underestimated what they were up against, whether it was the terrain or the enemy. They were people who closed their minds to other voices or ideas. 'Ego weakness and authoritarianism underlie most military ineptitude,' wrote Dixon. You could probably say the same about people who preside over most failing organizations and who seek out people who protect their illusions for them. There in a nutshell is the tragedy of so much modern organization, and especially those organizations that we rely on most.

The Nelson and Tryon management styles are something of caricatures. The patterns are sometimes discussed in management schools as Theory X and Theory Y. But there is one aspect of this that this book is specifically concerned with, and it is the element of human beings and relationships between them and how you manage to avoid that military incompetence in the organizations we all depend on.

The problem is that, in the early 20th century, when there is a problem – personally or organizationally – we tend to reach for an 'app'. We look for some kind of machine or system that can make it happen for us, avoid mistakes and build relationships – to make our workforces confident, loyal, committed enough to care strongly about their organization getting the best results. And here lies the problem. Because 'applications' can't do these things nearly as effectively as other people can. Systems don't build the relationships that Nelson had with his captains, or those that make things happen in everyday life. The argument of this book is that we are relying on systems more and more, and increasingly removing the human element that actually makes change possible. We are spending huge sums creating towns, businesses, governments, services which just don't work. They are not humane nor efficient, and nor are they sustainable. We have taken a wrong turning and we need to get back onto the road.

The truth is that we are all pretty familiar with the idea that you need to give more power and responsibility to frontline staff. There is hardly an employer out there that doesn't use the rhetoric of staff empowerment. We have also got used to the idea that IT is all about giving us more responsibility, and we all feel a little more powerful whenever we switch our laptops on. We really believe this. Yet, something has gone wrong.

The trouble is that most organizations we deal with might well believe their own rhetoric, but they hardly act on it. Most of our interaction with public services or big corporations is a frustrating business of squeezing something out of implacable and exhausting systems. Most of these organizations attempt to minimize human interaction, and are built firmly on the foundations of Admiral Tryon rather than Admiral Nelson. Most governments still have their sweaty fingers gripping desperately to the throats of their most distant employee. And, sure enough, things are grinding slowly to a halt, disasters are not anticipated, intractable problems are unaddressed and we sometimes feel – rightly or wrongly, after a bad day on the trains – that absolutely nothing works at all.

Why don't managers and policy-makers act on their own rhetoric? Why do we put up with the constant reminder of clunky, failing organizations all around us? Partly, of course, it is because the basic job usually gets done – you *can* go to the doctor and get treated, the food *is* on the supermarket shelves. Not everything goes wrong, as I have to remind myself after another exhausting encounter with Virgin Trains. But there is another reason too. It is because the stakes are so much higher, especially when IT combines with human error. Our corporations and services are suffering from an advanced case of what George Orwell called *doublethink* – the ability to believe two completely contradictory things at the same time. They believe they should be empowering their staff to use their human skills, but they are also terrified of what those same people can do once they are in front of a computer keyboard.

Consequently, we are building an edifice based on this fear of people. Our masters and managers believe that systems are the antidote to those messy, incompetent, accident-prone, unmeasurable, uncontrollable human beings. They are therefore engaged in producing a series of interlocking systems, which we interact with every day – public and private, government and business – that might as well have been designed to fail. Add in the amazing progress of IT and – bingo! – we have a tyrannical system that can construct an objective and bend people to fit it. But, most important, we have organizations that fail to do the job.

Human history has witnessed these trends before, in the last days of the Roman empire or the centralizing obsession of the Habsburgs, but technology has added the fantasy that human beings are tolerated in these systems simply while we wait for the software capable of replacing them. The Romans just wanted to administer their empire. These days, we hand over the business of feeding ourselves, clothing ourselves, managing our money and nearly every other function of life to corporate systems that talk to us via call centres and websites. We pay huge sums for the government to educate us, tackle crime and keep us well, to huge systems, managed by silos full of human drones, which suffer from precisely these same problems – and we wonder why it gets more expensive and less effective.

Although let's not dismiss the idea of human error. There are some crackpots out there, and everyone has their moments of madness. People can make a monumental mess of things. Take the case of Juan Pablo Dávila. In September 1994, Dávila was ensconced in his office in the financial district at the headquarters of the state-owned copper company, Corporación Nacional del Cobre de Chile, known to its friends as Codelco, then as now the biggest copper producer in the world.

Codelco supplements its mining with a little light trading on the world markets in precious metals and, back in 1994, Dávila was the head trader, actually the only trader allowed to do so electronically. He was about to prove, once again, the eternal truth of what has become known as Sod's Law, that if something can go wrong, sooner or later it *will* go wrong, and especially when that error gets compounded by IT.

It was a hot day and perhaps his mind was elsewhere when he pressed the 'buy' key absent-mindedly when he meant to press the 'sell' key to sell off some futures contracts on the London Metal Exchange. When he realized what he had done, Dávila was horrified to find that he had lost $30 million. According to the company guidelines, he was allowed to lose no more than $1 million, but nobody was watching, so he set about trying to undo the loss by shorting a million tons of copper contracts, effectively betting that the value of copper would go down. He also managed a little speculation in gold and silver on the New York Commodity Exchange.

Four months later, in January 1995, copper futures had risen in value by ten per cent and the accountants preparing the annual figures found that Codelco had lost $164 million in copper, $31 million in silver and $12 million in gold, apparently through a whole range of 23 brokers around the world. Dávila was sacked and, for a while, there was some doubt about whether Codelco could survive without defaulting on its debts. By the end, the whole incident cost Codelco about 0.5 per cent of Chile's entire gross domestic product (GDP).

These figures are not so impressive if you compare them to the staggering sums lost by most of the investment banks on Wall Street and the City of London more recently, but that was the result of collective errors of judgement stretching back decades. Dávila's mistake had been individual and instantaneous. Just one wrong button. His name even entered the language in Chile – 'davilar': to screw up massively.

Traders are horribly vulnerable to this kind of mistake. There was the sad story of the Salomon Brothers trader in New York, whose name was never revealed, who was the subject of a massive investigation by security consultants Kroll in 1999 because he had inexplicably made 145 sell orders for French ten-year bond futures. They discovered he had accidentally sat on his keyboard.

That is human error. We know all about it in our own lives, dropping the car keys down a drain, losing your credit cards in the fridge (I've done that one), absently forgetting your child's name just as you arrive to pick them up from nursery school. It is sometimes a relief that professionals do it just as magnificently as we do. When the pioneer aviator Douglas Corrigan turned the wrong way when he set off from Brooklyn in 1938 to fly single-handedly to California, and landed in Ireland instead, we cheer a little. When, in 2008, we heard that a government department had been sending £2.5 million in grants to Newcastle-under-Lyme under the mistaken impression that it was Newcastle upon Tyne, we cheer again and thank goodness it wasn't us.

The story of failure until our own day has always been the story of human error. The thesis of this book is that human genius is also the story of success, so the failure that we are seeing on a grand scale is more to do with the stifling of this genius than with the toleration of human stupidity – because human societies work best when they work collaboratively and when they allow for human brilliance, and that human brilliance may be more widespread than we think.

It may be impossible to work out who the individual mistakes are caused by, any more than you can isolate the individual human who caused those hideous military disasters – Napoleon's attack on Russia or the Battle of the Somme. But there were undoubtedly human beings involved, often lots of them. In fact, from the fall of the Roman Empire to the sinking of the *Titanic*, the story of failure has been the story of human error. Cock-ups, like the poor, are always with us and always will be. But over the past century – and largely because of the systems built by the originator of time and motion study, Frederick Winslow Taylor (a man who died winding his own watch), and his successors – that all began to change. Along with the panoply of time and motion study, which ironed out human difference in industrial and business processes, there emerged a new belief that, if people made mistakes, there must be systems for avoiding them in the future.

The traditional approach to dealing with human error has been to systematize: to take human decisions as far as possible out of the equation. The horrors of Chernobyl and Bhopal seemed like reason enough. You can always find a scapegoat, after all. And when you can make enormous savings by taking human beings out of the budget as well, then all the better. Human beings are expensive to employ: you have to pay them, provide them with desks and heating and insurance. You have to train, mentor and reassure them. You have to be nice to them – too exhausting for some of us – and deal with their sick leave. But the real problem is that they are fallible. Get rid of the human element and you get rid of the mistakes. Or so our organizations came to believe.

We have had just over a century of this process. When we speak to organizations we deal with, we usually get through to a computer these days. Even with those companies that advertise themselves as offering 'real human beings' – and there are a number of companies that do so (especially in the USA: Allstate and Geico Direct) – you still get through to a machine that gives you a whole range of options long before you get a whiff of anything human. When we manage to break out of call centre purgatory and get through to a person, they are operating software and seem strangely unable to deal with anything that is not included on their screen.

Every profession, including medicine and teaching, has been so constrained by systems and procedures until the point when it seems natural to discuss doing away with the human element completely. American IT prophets are enthusiastic about the prospects of a world where all teaching and most medicine is virtual, and where even virtual sex can be 'better than the real thing'.

This frustration with the way real people mess up is particularly prevalent among IT professionals. They use the term RTFM – an acronym meaning Read The Fucking Manual – and raise their eyes skywards when people fail even to do that, forgetting that their manuals are written by other IT professionals and are packed full of incomprehensible jargon, useless indexes and the kind of grammar that requires a pneumatic drill.

The hideous propensity of people to screw up was confirmed in one famous experiment by IT researcher Aaron Brown, who worked for the computer giant IBM. In 2004, he recruited five people, trained them how to replace a failed disk and gave them printed step-by-step instructions, then set them to work. Then suddenly the poor guinea pigs were given a simulated problem and Brown watched fascinated while – even without the stress of alarms going off or angry customers shouting at them – ten per cent of the repairs were done so badly that the data got lost. 'We are left to conclude that even the best-intentioned reliability technologies … can become impotent in the face of the human capacity for error,' he wrote, despairingly.

There we are: the implacable face of human error. No wonder managers have been doing their level best to remove people from processes, while at the same time nodding at all the rhetoric about empowerment. You would think, then, that we would now live in a largely fault-free world. You might imagine we could be reasonably certain, as we walk down the street, that the organizations we rely on each day will work smoothly and calmly. Some of them do, but most of them don't.

So there lies the same dilemma. We live in a world which believes two contradictory things at the same time: we must set people free to innovate to make things work, but we must control them tightly to make sure they don't make a hash of it. We believe both of these even though we can see the consequences all around us, and even though they cancel each other out in practice. We believe in people's skills but we are terrified of their ability to mess up. It's a perfectly sane position – both thoughts are true – but, in practice, this is a destructive contradiction which simply stops things working as they could. It is the reason for so much frustration.

Worse than that. As the rhetoric about leadership and responsibility ratchets up another notch, then so does the control. We invent ever more sophisticated systems to run our public services, usually at great expense. We talk about localization and responsibility and 'personalizing' public services, then – as soon as something goes wrong – we get tied up in such a mixture of regulation, jargon, targets, safeguards and systems that most of us lose the will to live. Then, far away and usually in the Far East, we outsource factories to make our trousers or mobile phones. The conditions in these factories are extremely controlling, so much so that, in the case of the Apple factory in Shenzhen, China, 12 workers committed suicide within six months in 2010.

Nobody wants more mistakes or more risk, heaven knows, so – despite the rhetoric – the systems win, time and time again. They win also because they are broadly supported by both sides of the political divide. The Right approves of systems because they believe people are basically untrustworthy. The Left approves because they see good process – especially in recruitment – as a way to root out racism, sexism and anything else that capricious bosses may reveal if they are not watched. Process seems to make decisions more transparent. Actually, it doesn't – we all know people manipulate the recruitment process to make sure the person they thought was best actually gets the job. Nor does it eliminate mistakes, as we all know to our cost – quite the reverse – but it *seems* to. It promises to, and those who run the world believe it.

Of course human beings play some role in mistakes. Captain Foley might have got it wrong, after all. Nelson's battle fleet might have been marooned on a sandbank. The irony is that the software and the processes that exclude the human element are also designed and managed by human beings, and they make mistakes too – they are just amplified. Bureaucracies are systems run by humans, but which paradoxically try to exclude human attributes, imagination, judgement, or their reverse. They are systems that are so complex that humans cannot control them. It makes them inhuman in their own very special way, rather as the Mediterranean Fleet under Admiral Tryon was able to hurtle to destruction with more enthusiasm and energy than any one individual could possibly have caused. That is the world we are in danger of plunging into when we try to exclude the human element altogether.

Inhuman systems also try to remake the world as if they *were* effective, and that can be painful for everyone. Take Croydon lorry driver Biorel Leon as an example. In 2007, he was told he couldn't have an Orange mobile phone contract by his local phone shop, yet for a year afterwards he was sent monthly bills. Neither the phone company nor the shop accepted that it was their fault, and demanded that he cancelled his contract – which they had refused to give him – and for which they would charge another £115. Next thing, the bailiffs are on the way round. What does he do in a world where there is suddenly nobody at the controls?

Or take the decision by Next to delete three months of callback requests by customers to reduce their backlog of customer service problems. Or Homecall, which changed the name of one customer without telling them, and then refused to speak to them on the grounds that they were not the 'named person'. Organizations which get big enough behave much like dinosaurs, incapable of ordinary human emotion, but quite capable of making a hefty meal of their own tails.

They may all be run by human beings, but the great bureaucracies – the profit-making ones and the public sector ones – don't behave in a human way. They are managed by a distant cadre of directors, who often believe that the existence and smooth running of the organization comes before anything else. In the case of many corporations, they sometimes believe their continued success somehow overrides morality, or that their only moral imperative is to raise the share price for their shareholders. This is an adolescent fantasy if ever there was one. We take our life in our hands when we contact these behemoths, as we have to do because they dominate so many aspects of our lives.

This is the new process-driven, non-human world – where benefits claimants have to hold on so long at the call centre on their pay-as-you-go mobile phones, just to make a claim, that it costs them £35. Where A&E nurses know what the patients in front of them need so urgently, but have to go through 20 pages of questions on their computer screens for each one. Where, three months after opening, Heathrow Terminal 5 was still losing 22,000 items of luggage, and these were criss-crossing the Atlantic in jumbo jets full of nothing else. This is a world where, all too often, we end up as hopeless supplicants or pathetic victims of large organizations which can only deal with us if we fit neatly into their systems.

Nobody wants to live in a world without processes. We don't want to make it all up again from moment to moment. But the truth is that it is this very distrust of the human factor by the technocratic systems that run our lives that is the real problem. It is true that humans make mistakes every day, especially when they are tired or unhappy. They don't fit neatly into the required categories. But they also possess the most extraordinary abilities, and do so from birth. They learn to live, to bring up children, to lead – most of them – generally happy lives where they fall in love and make things happen, and they do so every day.

Human beings do cause mistakes, but they also deal with human complexity better than any machine. We are in danger of forgetting one of the most important truths about the world: the human element may be a source of error, but it is also the only source of genuine change.

Still, something is in the air. There is a sense of public revolt against some of the systems that constrain us. Politicians and celebrities refused to submit to the new databases. There is a growing frustration about the sheer intractability of change and a growing suspicion that it isn't systems or processes or safeguards that make the difference between success and failure in human endeavour, but people.

There is even a suspicion that, despite all the investment in IT and organization we have seen, we live with the same old problems we always have done. Why are we still addicted to oil and petrol despite the disastrous consequences? Why, nearly 70 years after the Beveridge Report, are his Giant Evils –want, disease, idleness, ignorance and squalor – still so much with us? Why did teenage pregnancies go up despite the UK government spending up to £100 million over a decade to prevent them? Yes, and while we're about it, why do so few of the public clocks tell the right time and why do so few train lavatories have water in their taps?

The answer, as people are beginning to realize, is because so many of our organizations are great lumbering dinosaurs, with their human elements sucked out of them – huge humming machines with marble porticos, dedicated to their own continued existence, which could be atomized tomorrow without anybody noticing they had gone.

There is a growing understanding, not that people are infallible (we know all too well that isn't so), or that they are endlessly trustworthy and benevolent (they're not, heaven knows). It is that they are also what makes change possible. So if we are going to make our systems work again – we need to rediscover this simple truth, which provides the human dimension to the intractable problems that face us in the 21st century:

> If you employ imaginative and effective people, especially on the frontline, and give them the freedom to innovate, they will succeed. If you don't, they will fail.

This is what I call the People Principle. It is true no matter what the objective is, in business, charity or government. It is so simple that it is almost obvious, but it is also revolutionary.

The People Principle draws on quite familiar ideas. The pioneer sociologist Max Weber warned against the dangers of bureaucracy more than a century ago. The business guru Tom Peters has been aggressively putting the case for setting employees free for almost a generation. The bookshelves are stacked with titles and the business directories filled with consultancies urging the same message of empowerment. The point not only that this hasn't happened, but also – especially in the services we all use – we are drifting further and further away from it.

The point about the People Principle is that there is a new way forward. It isn't based on compromise. It is a radical rediscovery of the power of face-to-face human relationships, so that we might start solving problems rather than just hopelessly managing them occasionally. The research is also mounting that driving out the human element can be hugely destructive, especially in public services. This book draws on this research and tells some of the stories from what is an increasingly radical front line, to set out a manifesto for a new human revolution. The ten rules that follow will revolutionize the effectiveness of any organization, public or private, and are already doing so in the most advanced ones.

They will do so by injecting that crucial human dimension back into organizations. If we can make the People Principle work, we might have some chance finally – not just of getting to work on time – but of solving the great intractable social issues of the age, and doing so without bankrupting ourselves. When Margaret Thatcher's guru Sir Keith Joseph was first appointed to the cabinet in 1962, he famously remarked that he had been trying to get his hands on the levers of power all his life, but now he had finally clutched them, they didn't seem to be connected to anything. The human element is about finally finding a way of connecting the levers. Nor is it just the kind of naive manifesto that we all read with a forlorn, sinking feeling that it is based on dreams. This one is going to happen. In fact, it is starting to happen already.

The starting point is the amazing ability human beings have within themselves – some of them to a tremendous extent – to make things happen around them. So set out with me, if you will, in a school at the heart of post-industrial Stoke-on-Trent with a serious discipline problem – and that was just with the parents.

Find out more

The main starting point for writing this was my own book *Authenticity: Brands, Fakes, Spin and the Lust for Real Life* (HarperCollins, London, 2004), which concluded that the rising demand for something real in all areas of life tends to mean something with a human component to it and rooted to a human reality, and which I modestly recommend. You can get an American business perspective to this in *Authenticity: What Consumers Really Want* (James H. Gilmore and B. Joseph Pine II, Harvard Business School Press, Boston, 2007). The work of the Relationships Foundation also lies behind a great deal of this thinking (www.relationshipsfoundation.org). Their book *Building a Relational Society* (Nicola Baker (ed), Arena, Ashgate, 1996) is a good starting point. See also *The R Factor* (Michael Schluter and David Lee, Hodder & Stoughton, London, 1993).

Another useful starting point on failure might be an article in *Wired* ('A century of spectacular failure', Leander Kahney, 29 December 1999) or indeed Stephen Pile's bestseller *Book of Heroic Failures: Official Handbook of the Not Terribly Good Club of Great Britain* (Futura, London, 1979). Another perspective on failure, from an economics point of view, is Paul Ormerod's *Why Most Things Fail* (Faber, London, 2005), which is a fascinating study that I thoroughly recommend. But no book on failure could omit a reference to the classic text, *Parkinson's Law* (C. Northcote Parkinson, John Murray, London, 1958), which was expanded from an article in *The Economist* three years earlier. For brevity, clarity and brilliance, you can't beat it.

I also learned a great deal from Norman Dixon's classic *On the Psychology of Military Incompetence* (Jonathan Cape, London, 1976). The insightful management columns in *The Observer* by Simon Caulkin (www.simoncaulkin.com) have always been inspiring, and he has been key to putting the People Principle on the map, even if he wouldn't quite put it that way himself.

Recruit staff for their personality not their qualifications

We sometimes encounter people, even perfect strangers, who begin to interest us at first sight, somehow suddenly, all at once, before a word has been spoken.

(Fyodor Dostoyevsky)

Things that succeed have a personality behind them.
(Pat Brown, chief executive of Central London Partnership)

Summary

- Super-catalysts are people who are brilliant at dealing with other people. Their relative absence in our public services is one explanation of why they are sometimes so intractable.
- Working imaginatively with people who can make a difference is exciting in a way that watching over processes is not.
- The important thing to remember about human catalysts is that we were all born with the necessary skills.

You can always recognize a failing school. The mini-class of naughty pupils doing their work in the headteacher's study. The coats all over the floor. The bored faces of the children sitting at their desks with the sun outside. I went into one primary school recently where the headteacher said: 'You feel like weeping at the end of the day.' If ever there was a sign of something not working, the abject misery of the person in charge is probably it. So when anyone can turn round a failing school, it has to be a clue about making other systems work too.

So take a journey with me for a moment to one school which turned around: Mitchell High School, right in the middle of an impoverished but energetic housing estate on the outskirts of Stoke-on-Trent. Back in 2001, one senior member of staff faced down a furious local parent, who was waving a pair of muddy shorts at him, and was smacked around the face with them. Three years on, an amazing transformation had taken place. The angry parent was a member of staff, as were many of her friends, and – here's the strange bit – they were paid partly in chocolate coins.

This unusual approach to turning round a big secondary school came at a time when other schools were putting in extra cameras and security gates just to avoid violent confrontations like that. The fact that Mitchell went a different way, and became such a success story – as it did – was largely down to the new headteacher, Debbie Morrison (called Debbie Sanderson in those days). A closer look at the way she works reveals a great deal about the human element and why it is so important. Because if you hang around successes for a long time, you realize that one of the things they have in common is the people, like her, who are at the heart of it and who seem to be able to make a difference when everything else seems to point the other way. I call them super-catalysts, and it doesn't have anything to do with qualifications.

When Debbie Morrison took over, immediately after the shorts incident, a fifth of 16-year-olds left Mitchell High School without any qualifications. A few years later, that figure was down to four per cent and two thirds of the school were getting five or more GCSEs at grades A–C. You might think that the basis for this success was some kind of innovative government programme that could be rolled out anywhere. There were government programmes, of course – there always are – but the truth is that this didn't really apply to Mitchell High School. The key to Mitchell's success was partly the very human skills of the headteacher, but taken to a whole new level. In was, in fact, a dramatic demonstration of how people have skills that go some way beyond government programmes, and it had knock-on effects in the local community too.

Debbie Morrison may have been a born teacher, but she actually trained as a chartered accountant. In a disaffected moment, she saw an advertisement for a teaching job in a local school in Derbyshire and applied for it. She rose quickly through the ranks until she arrived at Mitchell, thrilled by what she describes as 'the life and dynamism' there. But behind that buzz, the neighbourhood was not exactly going places. 'It was a highly depressed neighbourhood with a self-limiting belief system,' she says. 'They really didn't believe it could be any better, across the whole community. It had become self-fulfilling. There was real aggression and real disaffection there, and a kind of acceptance that they had no influence over their own futures. Everything was always done *to* them.'

Worse, when she arrived at the school for the first time, Debbie was warned not to walk along the corridors on her own. Local parents told her that she had no chance of sorting things out, because two strapping men had failed before her and she was just a young woman. Nor were the parents just apathetic about learning – some of them were downright hostile. Only a week into her job, as she locked the school gates, one parent holding a spade threatened to kick her 'fucking head off'.

Her strategy was to engage with the most vociferous local mothers, not just to listen to them but to ask them for help. She pursued one of the loudest from the local estate, a mother of two particularly challenging twin boys. 'We kept phoning her up,' she says. 'We kept asking her to come in, and I thought, if I could get Donna on board, I could get the whole community.' Some people might not have the human skills to engage in this way, possibly even most people. But within a few months, Donna had enrolled alongside the children in the health and social care course. This was a major success in itself. Her mere presence in the classroom gave a message to the other children that learning was respectable, and she was also able to tell them some first-hand stories about childcare too.

The next step was to employ Donna for an hour a day during the lunch break to stand at the school gate and take the names of any children leaving for the afternoon wandering round the shops. Some of the staff were nervous about rubbing shoulders with a parent in the staff room, especially such a difficult one – even sharing the staff toilets with them. But Debbie knew the strategy was working when one of the other local mothers took her aside and said: 'I want to do what Donna's done.'

'You've given me a bite of the apple,' said another one later. 'Now I want the whole apple.'

Others came in as classroom assistants, but Debbie refused to accept the usual boundaries between the school and the outside neighbourhood. There were self-esteem classes for adults and courses on basic literacy. Other local parents were employed in various outreach work. There were lots of award ceremonies, prizes and awards afternoons. There were grants to take over empty buildings near the school. 'I tried to find something that somebody was good at and build on it,' says Debbie. 'When I managed to find space at a minimal rent, I had somebody in mind and said to them: "If we put ten computers up there, can you manage that house in the afternoons?"'

In her new school, Coundon Court in Coventry, she has launched an ambitious project to train pupils, teachers and parents in life-coaching skills, beginning with an intensive course in the summer holidays for

350 children, chosen – not because they were the brightest and best – but because they most wanted to take part and give something back. Fourteen went on for a four-day training course, covering everything from emotional intelligence to ethics. Even before the end of the course, it was clear that 12-year-olds could be brilliant coaches of 50-year-olds. One of the pupils is now coaching Debbie to organize a better work-life balance, go to the gym and drink enough water, and is sending emails to remind her of the goals she's set herself.

The chocolate coins began as a thank you gesture at Mitchell, and they are a clue to the meaning of this story. 'OK, it was cheesy but it worked in that context,' she says. But how did it work? They are not exactly a payment: you can't put them in the bank and they tend to melt if you keep them in your pocket. They have no value, and yet Donna valued them so much that she still keeps some of hers, locked away in a cupboard. They are a delicate balance as well. You can hardly imagine government guidelines for payments with chocolate coins. Anybody else shelling out chocolate coins might just irritate people, or worse. But Debbie managed to pull it off. The chocolate coins worked, not just because they were an informal, human touch, but because they were very personal.

What seems to me to be amazing about Debbie's success at Mitchell High School is that she had no obvious plan in mind when she set out. 'I had a vision of self-supporting communities and a sense of how it could feel, but I didn't really know how I was going to get there,' she says. 'It was like lottery balls flying in the air. It was about jumping from ball to ball, riding on the energy.'

Much of the success came from talking up the school and the community, helping pupils and parents realize they could achieve things if they set out to. But by itself, that's just spin. The point is that Debbie Morrison is a brilliant example of someone using their human skills to dramatic effect, cajoling, challenging, comforting, imagining. You could never boil down how she works into some kind of computer programme, still less deliver it virtually. She knows instinctively how people work, and forges relationships with them to make the change happen.

Of course, there are lots of people like her, though they are rarely given the credit they deserve. They succeed because they can look at other people and see potential in them that officials and institutions don't see. They don't just re-categorize them, they engage in such a way that change takes place. This isn't just a matter of leadership – there are leaders out there who don't know where to start when someone is in front of them. General Montgomery was a disaster on a one-to-one basis. These are people who make things happen instinctively, often in tiny ways, and do so brilliantly. That is what makes them super-catalysts.

When I first ran across Debbie Morrison, it was at a government conference about 'extended schools', for which Mitchell provided a blueprint. She gave the keynote address, packed with inspiring stories, which made you believe that anything might be possible. The next speaker was the civil servant charged with rolling out extended schools across one of the English regions and, within a minute or so of him beginning to speak, it was pretty obvious that he would fail. He was revealing the besetting sin of officials, which is to boil down successful examples to universal principles which they believe can be applied anywhere.

This is how the process goes. First, they take an intractable problem about neighbourhoods, communities and places, then they remove all of what they see as the dull and mundane but essentially human details.

Second, they formulate some abstract maxims that can apply to any situation or any community.

Third, they appoint somebody who can be trusted to put those maxims into effect without taking any notice of local peculiarities.

Fourth, they assign a narrow measure to every aspect of the task, and convince themselves that you can somehow capture and pin down the progress by measuring it.

The trouble is that you can't actually separate the general from the specific. It is the little things that matter most – the looks exchanged between neighbours, the small repairs to minor pieces of vandalism – that will make the difference between success and failure in a neighbourhood, the human encounters that make such a difference in the school. Rolling it out like this pretends that the people aren't crucial. It imagines that it can all work fine if individual relationships aren't forged. The thing about super-catalysts is that they turn their visions into reality in their own way, not according to the boiled down maxims preferred by those who employ them, but using their human skills in the way they know best. This can make them the objects of suspicion in hierarchies which have become too divorced from the way people actually behave.

Debbie Morrison revealed that the head of the coaching programme they brought into Coventry called her a 'deviant', in an admiring kind of way. She smiles at this. 'Unless you have a perception of the norm, you can't really deviate from it,' she says. But there is no doubt that others like her, particularly in public services, are seen as exactly that by the authorities, especially when they dealing directly with the public.

If care staff help disabled people back into wheelchairs when they fall out, and they are not insured to do so, they often get reprimanded. One local government officer told *The Guardian* in 2008 about a boy with learning difficulties who loved swimming, especially with his brother. But the brother didn't have learning difficulties and the swimming group was only for those who did, so he was barred for 'health and safety reasons'. You have to be a deviant, at least a bit of a dissident, to break rules like that and stay a little bit human. It may even be this very ability to break rules which makes all the difference between success and failure.

A decade on, Mitchell is facing closure again. But what is most important about its renaissance under Debbie Morrison is that, if the human element had been removed, in case staff didn't follow the correct procedure or because the risks to children were too high – or maybe because a computer programme was considered more efficient – then Debbie could never have done her job. This matters enormously because it is at least a clue to the key question: why do things go wrong so often? Why, after huge investment in systems and the latest management consultancy, are so many schools still failing? One answer is that the super-catalysts are either missing or their skills have been deliberately frustrated.

It is rather an urgent question. Businesses call people like that 'entrepreneurs'. There are even 'social entrepreneurs' in the voluntary sector, but there is no equivalent word in the public sector. In fact, people who work imaginatively or emotionally are occasionally objects of suspicion in public services, especially when they are manning the phones or looking after children. Even in business, there are moments when magnetic entrepreneurs are not quite as welcome as you might expect, especially when they are in a hierarchical system that tries to control their script and reaction.

No, being an entrepreneur is usually about generating money. That is a related but different skill. Super-catalysts are people who are brilliant at dealing with other people. They are often leaders, but they are not just leaders. Of course, they are often entrepreneurs, but they are not just entrepreneurs either. And their absence in our public services is at one explanation of why they are sometimes so intractable.

It has taken a long time for policy-makers to re-discover the crucial importance of human relationships in education, but they have done so faster than most other areas. The US teaching consultant Doug Lemov had been a big advocate of data-driven programmes and standard tests to improve schools, but he changed his mind after a particularly depressing visit to a failing school in Syracuse in New York state.

They seemed to be doing all the right things. They had caring teachers and small classes, and sophisticated software to analyse every pupil's test results, but the classrooms were a disaster. Lemov watched while a teacher launched into a long debate with a pupil about why he didn't have a pencil. Driving home afterwards, Lemov reflected that – despite his battery of techniques, software and analysis – he had little idea how to help schools actually teach.

His confusion coincided with a series of studies in the USA into the success of President George W. Bush's No Child Left Behind programme. When they looked at the statistics regarding the huge number of factors that schools have some kind of power over, all the usual ones showed just a tiny impact. What really made the difference was which teacher the pupil had been given. There were huge gaps between the achievements of pupils when everything was exactly the same – the same curriculum, same school, same background – but different teachers. Parents obsess about the right school to send their children to, but the latest research suggests that actually the school is only important because it houses the teachers.

It is only common sense that individual teachers make a difference, but these findings were a shock to the teaching establishment, largely because it implied there was something indefinable about individual teachers that couldn't be measured. One leading education policy analysts told the *New York Times magazine* that it was 'voodoo'.

The accepted solution now, at least in the USA, is to give teachers performance-related pay, as if somehow low standards was the result of them not trying hard enough. Lemov didn't think that was the best way forward and set out to define what made a good teacher, travelling around the country filming the most successful teachers at work. The result was a set of techniques that came to be known informally as 'Lemov's Taxonomy'. They are little things, such as standing still when you give instructions, that Lemov believes can turn ordinary people into brilliant teachers.

The charity Teach for America was also puzzled about their research findings. Their teachers were doing good work, but a handful of them were succeeding way above the others. They asked a philosophy graduate called Steven Farr, their head of research, to find out why. Starting in 2002, he tracked down teachers that were making a big difference and found they had a number of things in common: they constantly re-evaluated what they were doing, they obsessively recruited the families of their pupils into the process and they battled bureaucracies that were getting in their way – they were 'deviants', like Debbie Morrison, in that respect. But by the end of the research, two things stood out above all others as a measure of successful teachers – and it wasn't background or academic accomplishments. It was perseverance and determination, but it was also satisfaction with their own lives. These were secure people, at ease with themselves. Teach for America now tries to identify people who can show a track record of continuing to try when they recruit them.

The implication of this – and of the work done by academics in the same field, on both sides of the Atlantic – is that you can learn to be a super-catalyst – at least in the classroom. The sum total of all Lemov's 49 techniques would be a great teacher. That is why philanthropists such as Bill Gates have got involved in finding out what makes teachers brilliant. It is why the Obama administration decided to double the funding for teacher training. It may be that being a catalyst is something that you can learn. It isn't clear what the balance is between learning and genetics in the super-catalysts. Perhaps it doesn't matter. The real point is that success or failure is primarily down to individual human beings. That is the human element.

When *The New York Times* devoted an issue of their magazine to the question of whether individual teachers are important, they described Lemov at the back of a class watching a really brilliant educator at work. 'You could change the world with a first-year teacher like that,' he said afterwards. People make a big difference – so much so, that in a world of systems and processes, it looks as though they could create serious change.

The man who has emerged as one of the most successful advocates of recruiting by personality in the UK is John Timpson, the maverick chairman of the high street shoe repair and key-cutting chain that bears his name. He organized the management buy-out of his family firm from the Hansen Trust in 1983. Fifteen years later, they began to face nose-to-nose competition from Mister Minit. Timpson had a disturbing interview with the owners – the Swiss bank UBS Capital, a rival with bottomless pockets – where they told him they made a speciality of swallowing up family firms.

Chastened by this threat, he struggled with the question of how to fight back and realized they had only one asset available to them to compare to the UBS billions: their shop managers. But how could they use them better? The answer Timpson came up with was to give them more flexibility to adapt to the local situation, but not just that – he decided to give them total control, and to trust them to get on with it.

Timpson called this new doctrine 'Upside Down Management'. He wrote about it widely in the staff publications and urged it every moment he could. Nothing much changed. He realized he would need a symbolic gesture to make staff understand: he gave each of them, however junior, a budget of £500 to settle complaints, and – much more radical, this one – he also let shop managers decide what prices to charge for their products.

Timpson Ltd had been started by his great-grandfather 140 years before, but never had any of predecessors flown so much in the face of conventional business wisdom. It still took three years to convince the staff he was serious, and even then there were huge implications to be faced, but it was a huge success. Timpson eventually took over the loss-making Mister Minit, and made it profitable, and at the same time grew the business from 200 to nearly 700 stores around the UK. He continues to make acquisitions and still has a tough time inculcating his management system in the new executives. Yet it is also a showcase for how super-catalysts can work in business when they are encouraged.

But the key to employing more super-catalysts is picking the right people. Timpson's magic decision was to stop recruiting staff on the basis of whether they could mend shoes or cut keys, which – as Timpson said – reduced their pool of potential employees to about 30,000 people in the UK, and start recruiting them on the basis of their personality. This is a theme that is repeated over and over again through those few organizations that genuinely understand that people are central to success, rather than just using the rhetoric. They have faith in their training to fit anyone for the specifics of the task, but they want the right people to begin with. They don't worry too much about qualifications or paperwork. Like the recruiters at Teach for America, they look for what the people *are*.

It seems such an obvious way of going about recruitment, yet so many businesses go about it in a completely different way. The vast majority of public servants, in this country and abroad, and those who work for large organizations, public and private, are recruited using formulae, given an equally formulaic training and then abandoned on the job. Lip service may be given to their ability to make things happen and their entrepreneurial flair, but that is not actually what is expected of most of them. People in big organizations tend to be expected to keep rules, not bend them. The fact that so many people do actually make things happen every day, especially in public services such as the NHS, is simply evidence of their potential to do more.

Some of the most successful companies in the world have shifted over to recruiting more super-catalysts. Procter & Gamble now looks for what they call 'in-touch capability', a fancy way of saying emotional intelligence. Southwest Airlines, one of the few profitable airlines in the world, deliberately now hires people on the basis of attitude and leadership, not on their technical skills or experience.

But the company that has thought most about recruiting differently and re-organizing the business around super-catalysts is the giant conglomerate General Electric (GE). When Jack Welch took over GE in 1981, one of the world's wealthiest companies, it was earning $1.5 billion a year and had around 400,000 employees. When he left 20 years later, it was earning $12 billion. The number of employees was slashed to around 299,000 by the end of 1985, earning Welch the nickname 'Neutron Jack', a reference to the bomb which kills people but leaves buildings standing. Welch was the son of a railway guard from Peabody, Massachusetts who joined GE as an executive in 1960. In fact, he had accepted another job quite early in his career at GE, but was taken out to dinner and persuaded to stay by a young executive called Reuben Gutoff, who promised to work with him to cut bureaucracy and create a small-company environment.

Welch set about transforming a corporate monolith that he believed was throttling itself in its own processes. That meant dumping the idea of strategies. Who reads them after they are written so laboriously? It meant abandoning the great edifice of corporate measurement. 'Too often, we measure everything and understand nothing,' he said. Under Welch, GE would just measure customer satisfaction, employee satisfaction and cash flow. It also meant, as far as possible in a huge company, striving for informality. It wasn't a small company, heaven knows, but Welch wanted it to feel like one.

None of this suggests that Welch was a leader to be copied in any other way than this – his environmental record was appalling. But he knew that the kind of people he needed would sometimes be different from the ones he had got. Take away the structures of bureaucracy and people look different. 'Now you see some of them wilt,' he said, about some senior executives after the process of 'de-layering'. 'That's the sad part of the job. Some who looked good in the big bureaucracy looked silly when you left them alone.'

That is what happens when large organizations transform themselves into effective ones. The people who made the wheels of bureaucracy go round, who felt more comfortable preventing useful activity, lost their role. Those who were held back from making relationships work by the bureaucracy are set free to set the place alight, but sometimes it is horribly obvious that many of your staff aren't suitable after all. So this is one of the keys to putting people back at the heart of organizations. It means making recruiting absolutely central, which is why Welch said he spent half his time recruiting the top positions in the company.

Two decades on since he took over, the Welch message has filtered down through corporate America. McDonald's CEO Jim Skinner now personally reviews the development of his top 200 managers. Welch's successor at GE, Jeff Immelt, meets his top 600 staff before they are appointed. Medtronic boss Bill Hawkins spends half his time dealing with the careers of his key managers. These are huge commitments of time, and there is little point in doing it if you are then going to constrain managers with processes and peer over their shoulder all the time once they are in the job. What's more, generally speaking, we don't do it like this in our public services, yet who is to say that hospitals and schools are any less important?

This all assumes that every staff appointment is going to work out, and – as managers wander around their domains – they are bound to find areas that are not really working. Of course, most staff will not attain super-catalyst status. Some will also actively frustrate the others. The real question is not how you control the organization to regulate this minority, but how you can help the rest learn and how you can exclude those who won't or don't. The challenge of Upside Down Management is the same as the one faced by people such as Jack Welch. It is not just getting hold of those super-catalysts in the first place, but how to deal with those who won't learn. There is a tough element to this, which Timpson typically expresses in stark terms. You need a policy of what he calls 'getting rid of drongos'.

Usually the drongos are obvious in interviews, he says. One brought his mother who answered all the questions, another listened to his head-phones throughout. But sometimes you don't, and then they have to go. Timpson always asks himself whether he could last through lunch with the person before him. If not, and he has inadvertently employed them, he recommends telling them as directly as possible, rather than pretending that this is a disciplinary matter and going through the motions to avoid action for unfair dismissal. He gives them a big pay-off, helps them find another job and breathes a sigh of relief.

Upside Down Management means putting training at the heart of the enterprise, as companies such as Eli Lilly and Nokia have done. It means mentoring and coaching have a central role. But it also means people who don't shape up have to go. Welch set up a particularly ferocious system that graded all GE staff every year as A, B or C. The As were promoted and the Cs sacked. 'Move them out early,' he advised. 'It's a contribution.' But then, as I said, not everything about Welch's reign over GE was admirable, and it isn't clear that terrifying your staff is any more effective.

Even so, this is the most uncomfortable aspect of the human element. It means people have to be easier to dismiss, not because there is a downturn or for reasons of economic expediency, but if they just can't connect. If they can't read or write, then everyone accepts that they can't do the job. If they can't relate to people, can't influence or make things happen, and no amount of training shifts their attitude, then they should not, for example, be teaching children or healing patients or serving customers. But if Teach for America and Doug Lemov are correct, and these vital skills can be taught, then the main question is how to teach them.

Unfortunately, and for many good reasons, we have come to believe in processes more than we believe in people. We have bought the argument that human beings are fallible, and worse than fallible – we are afraid they are racist, sexist, accident-prone brutes (not us, of course). We don't believe that recruitment or decisions of any kind can possibly work without the whole process being taken apart, set down and made visible. Many people, especially in the public sector, are actually hostile to the idea that people can make a difference. A friend of mine shocked a room full of health administrators by suggesting that different surgeries have a different atmosphere about them. They objected to the implication that the smooth system could be over-ridden by individuals.

Timpson has flown in the face of all this by launching a radical experiment to recruit his staff directly from jail, and is doing so already, with training workshops in Wandsworth and Liverpool prisons. This looks like he is testing Upside Down Management to destruction, but it is also pushing to try to recruit catalysts, no matter how they appear on paper. There are already a number of success stories – at least one of the former prisoners they employ is now a branch manager. But Timpson is constantly stymied by the prison officers who send him candidates because they have kept to the rules or kept their head down. 'We don't want that,' he says. 'We want personalities, who are hard-working and keen.'

The problem is that the system wants to systematize. Of course it will systematize the definition of the 'best' candidates. It is also true that trashing the system might make it easier for racists or morons to abuse their position in the recruitment process, but actually the system only gives the *illusion* of oversight. You do need some kind of transparency as a safeguard, but processes – because they are usually based on apparently objective measurement, which can't sum up people – only *seem* to provide it.

The alternative is that the organization slowly fills up with the wrong appointments, and they make more wrong appointments. 'If you recruit drongos, they will bring in more drongos,' says Timpson, 'because they feel comfortable with other ones around them.'

No, the solution has to be asserting that people have human capabilities which are the most important of all. If we are going to treat people as potential catalysts, it means we have to put our faith in them again, though not without oversight, and put aside our misplaced faith in processes. It means asserting that management, government and administration are about leadership, not about the manipulation of programmes, however visionary. 'Our whole belief now is that running things is about process, and that all you have to do is follow the process and all will be well,' says John Timpson. 'But if you recognize who are good people, you don't need processes, you can run the business much more efficiently, with fewer people, and you can have more fun.'

That is the key to it. Super-catalysts are life-enhancing. Working imaginatively with people who can make a difference is exciting in a way that watching over processes is not. Watching over systems is little better than machine minding. It's no fun.

Can you recognize super-catalysts in the street? I don't think so. They've certainly got energy, and instinctive sympathy, maybe even charm, but my experience of those people you run across who really transform situations around them is that you can't generalize about their personalities. Some have huge self-belief, but that can also get in the way of making relationships work. Some make things work in a very modest way.

What I have learned from researching this book is that super-catalysts are not strange and unusual people with a genius for relationships. They are not people with psychic gifts or knowledge that is shared by only a handful of people. This is not an assertion that can be proved one way or another, but I believe they are using skills that we all possess as part of being human. They just manage to exercise them in an intense way and, despite all the organizational barriers against them. They are awkward people. They don't accept defeat.

When I think back to all the examples of people who make things happen that I have seen – as we all have – it seems to me that situations are changed, not just by headteachers or chief officers or business leaders, or anyone with a reputation as a transformational person, but by very ordinary people in playgrounds, front rooms and front gardens. The important thing to remember about human catalysts is that we were all born with the necessary skills. We see them put into effect around us every day, in families and neighbourhoods. We need them to bring up children, make relationships work and make any kind of living. The fact that many of us make a mess of these tasks proves nothing; most of us manage it in the end, and many of us do so spectacularly.

That is what super-catalysts mean. They have managed to hone and transfer to the workplace the human skills that most of us already have. Most of us learn to live, to bring up children, to lead generally happy lives where we fall in love and make things happen all over again. This awesome individual genius is the key to the human element.

In these areas, so much *does* work. Children are socialized and turn out humane and imaginative. People look after their neighbours in their hundreds of thousands every day, despite the rhetoric about 'broken Britain'. There are super-catalysts everywhere, and the best way to make our organizations work is to recruit them, then train them and – as far as possible – to get out of their way.

Find out more

The term 'super-catalyst' is one of mine, so there isn't a huge amount of other literature on it. Of the main examples, Debbie Morrison is a key figure in the extended schools movement, and you can keep up with this through the newsletter *Extended Schools Update* (www.teachingexpertise. com). Most of the information here is from an interview I conducted with her at her new school in 2009. For the American discoveries about the importance of individual teachers see Elizabeth Green, 'Can good teaching be learned?', in *New York Times Magazine*, 7 March 2010; and Amanda Ripley, 'What makes a great teacher?' in *The Atlantic*, Jan/Feb 2010.

One of the key texts for this chapter is John Timpson's *How to Ride a Giraffe* (Caspian, London, 2008). The quotations from Jack Welch are largely from *29 Leadership Secrets from Jack Welch* (Robert Slater, McGraw-Hill, New York, 2003) and, more recently, *Jack Welch Speaks: Wit and Wisdom from the World's Greatest Business Leader* (Janet Lowe, John Wiley, Hoboken, 2008). There is more about how some of the biggest companies are concentrating on better recruitment in *Fortune* ('Leader machines', Geoff Colvin, 1 Oct 2007).

I also thoroughly recommend the recent book by Michael Fielding and Peter Moss on relationships in schools and why they are so vital (*Radical Education and the Common School: A Democratic Alternative*, Routledge, London, 2011).

Dump the rulebooks and targets

'You are to be in all things regulated and governed' said the gentleman, 'by fact. We hope to have, before long, a board of fact, composed of commissioners of fact, who will force the people to be a people of fact, and of nothing but fact. You must discard the word Fancy altogether.'
(Charles Dickens, *Hard Times*, 1860)

Use figures as little as you can. Remember your client doesn't like or want them, he wants brains. Think and act upon facts, truths and principles and regard figures only as things to express these, and so proceeding you are likely to become a great accountant and a credit to one of the truest and finest professions in the land.
(James Anyon, the first accountant in the USA, speaking to new recruits, 1912)

Summary

- However incompetent staff may be, they will always be skilful enough to make targets work for them rather than against them.
- You have to be there, asking questions, sniffing the atmosphere, finding out what is actually happening at the frontline, and preferably by yourself. That is what allows people to put their human abilities into effect.
- Managing doesn't mean abandoning control altogether. It means shifting from specifying every possible response to setting out broad principles.

If you follow Rule 1, and recruit super-catalysts – based on their personalities and ability to transform relationships – then it makes no sense to control every detail of the way they do their job. The rhetoric of empowered employees scorns the very idea. But we all know that is exactly what happens to most employees, especially in our public services. Take one energetic official chosen at random from the public sector in the UK, Richard Elliott, a former member of the Bristol drugs action team. As such, he had to keep his eyes on 44 different funding streams, nine different grids and 82 different objectives imposed on him by managers, funders and the government. Before he resigned, he reckoned that he and his colleagues spent less than 40 per cent of their time actually tackling drugs issues. But it is not so much the time constraints or the bureaucratic overload I am talking about here; it is the sheer complexity that constrained him.

Elliott compared his management regime to a kind of addiction on behalf of the obsessive and narrow measurement of his performance. 'Monitoring has become almost religious in status, as has centralized control,' he said. 'The demand for quick hits and early wins is driven by a central desire analogous to the instant gratification demands made by drug users themselves.'

You can't pretend that the private sector is very much better. Most employees have targets, and must report constantly on screeds of other numerical data which persuade managers at the top that they have a clear view of how the company is performing – it is the fantasy of the driving dashboard. But not all companies go down the targets route. John Timpson's Upside Down Management rejected everything that might narrow the relationships between customers and frontline staff, including budgets (largely fictional) and strategies (nobody ever looks at them again). He let staff make up their own job titles. He threw out the voicemail system for customer complaints but, most symbolically, he threw out the computerized EPOS (electric point of sale) tills which controlled staff and sent data to head office. From then on, they could charge what they thought was right.

No decision had quite the impact of rejecting the wired up tills. 'As soon as you stop trusting people, you put in a whole lot of systems that stop you serving customers,' he wrote. If staff were allowed to make relationships with customers, to enjoy communicating and to have fun, then everything else would fall into place, he said. When there was any doubt about the key they had just cut, he even encouraged staff to tell customers they could take it home, see if it worked, and pay tomorrow. Not something in the rulebook of Harvard Business School. In fact, Timpson positively relished his rejection of the rulebook:

> I love it when our branch colleagues use their initiative to do things that would be against most company rules. Some shops

have newspapers for customers to read; others provide sweets for children. Some even make customers a cup of tea. They are happy to look after customers' shopping. They let them use our loos. Most shopkeepers find an excuse to avoid this sort of free service, often blaming health and safety. We encourage colleagues to ignore imaginary red tape. They amaze customers by doing helpful things such as carrying shopping to the customer's car.

Rule 2 suggests that the vast majority of organizations – and especially public services – should follow Timpson and fling out the rulebooks, targets, specifications and standards that frustrate real face-to-face relationships at work. It is tough, and it seems terribly dangerous to the hierarchy, but – if we believe that human relationships lie at the heart of successful human endeavour – then it makes sense to get out of the way of people who can forge them.

Timpson went so far as junking the company's sales target system. Now they just compare each shop's takings with those of the previous week. They give frontline staff the flexibility to respond, trusting that they have recruited the right people and trained them effectively, and knowing that they can pick up trends quickly, and act on them, because they are in direct contact with people. The trend has been to try to build flexibility at the top of organizations, at the expense of ever tighter control at the bottom. Upside Down Management, as you would expect, is the reverse of that – and the very reverse of what is currently taught in business school. The flexibility is in the front line. It is allows front line staff to emulate Captain Foley's leap of faith, if they see the opportunity. Of course, it also makes them more responsible if the leap of faith goes wrong.

In fact, Timpson believes that, if he was ever to let management consultants loose on his company, they would start by taking all the power away from staff, would save £3 million immediately and 'in three years, the business would be completely screwed up'.

It remains a tough proposition for most managers, in the public or private sectors, though there are – as we shall see – companies which are travelling on related paths around the world. He described the process of initiating the managers of the photo-processing company which Timpson's acquired in 2009 in the theory of Upside Down Management. 'We have got to the stage when they have bought into the concept of letting their staff get on with the job,' he said. 'They want to give them freedom but they still want to do what they did before.'

It is easy to dismiss Timpson as a doyenne of the Grumpy Middle-Aged Men school of business – an atavistic rejection of the modern world – but he is too successful for that. He also took the conscious decision not to go public and have Timpson shares traded on the stock market. By doing so, he almost certainly foreswore a personal fortune, but he has retained the flexibility in return – aware that it meant handing control of the company over to analysts and traders with no interest in innovation which wasn't immediately profitable. He did so on the advice of his wife and he says it was the best decision he ever made.

Upside Down Management also fits clearly into an American tradition of maverick business success. When Timpson's son James took over as managing director in 1998, he visited the key corporate exponents of this in the USA, Southwest Airlines (managers carrying things for customers), Ritz-Carlton (budgets for junior staff to solve problems) and WL Gore (no job titles). The real question is not why Timpson has taken the turning it has, but why so few big British companies have followed in the imaginative American direction of empowering their staff (The Body Shop is one big exception that proved the rule, but which did become a public company, much to the regret of its founders).

Is it that business orthodoxy is less challenged here in the UK? Is it that we have so few banks that they exercise greater control? Is it somehow that the influence of the City of London is stronger over the UK than Wall Street is over the USA? Is it that companies and services are too terrified of being overwhelmed by their customers, by either the weight of their demand or the irritation of the extra involvement? 'Why do they still not get it?' Timpson asks.

The answer isn't clear but there is an even bigger question. It is why public sector organizations have been so slow to follow suit, especially when some of them began to do so in the entrepreneurial 1990s. Can you imagine an NHS outpost with posters up, as they are in Timpson shops, signed by the chairman himself, which say: 'The staff in this shop have my authority to do whatever they can to give you amazing service'? Can you imagine those notices in hospitals or post offices? Can you imagine probation officers set free to make things change for their clients? Or is it that the super-catalysts are just too scarce to make public services seriously customer-focused in this way?

'I don't see why you can't do that,' says John Timpson. 'But the first thing that would happen would be that you would get loads of people coming to you and saying why it won't work. Then you'll have lots of people telling you what a good idea it is but not actually doing it. It only works if the person at the top believes with a passion that it must happen.'

The idea of empowering staff has emerged during the rise of precisely the opposite idea – that every detail of what people do at work, and the results of what they achieve, needs to be measured and fed through to managers at the top.

Where did it come from, this obsession with targets? Some people date it back to the moment in 1903 when the time and motion study pioneer Frederick Winslow Taylor rose to his feet in Saratoga Springs to explain his idea that every factory could be measured to work in what he called 'the one best way'. Maybe it was actually James Oscar McKinsey, the first management consultant, whose consultancy still lives by the highly misleading maxim 'everything can be measured and what can be measured can be managed'. Maybe it was the technocrat's technocrat, Robert McNamara, who imposed 'kill quotas' on soldiers in the Vietnam War, only to find that the deaths rose but victory stayed elusive. Whatever it was, the management business has spawned a vast industry that churns out targets, specifications, standards and obscure acronyms, while an even bigger industry puts them into effect. The idea dominates consulting just as it now dominates government – the Blair government introduced 10,000 new numerical targets in their first term of office, on everything from vandalism to the state of sailor's teeth in the navy.

The side-effects of these targets have become much clearer, as managers tend to become experts in the business of manipulating the data. The problem is that controlling people with numbers never works. The principle that numerical measurements will always be inaccurate if they are used like this is now known as Goodhart's Law. The reason is that, however incompetent staff may be, they will always be skilful enough to make targets work for them rather than against them. Take for example, the rule that patients shouldn't be kept on hospital trolleys for more than four hours. In practice, some hospitals got round this by putting them in chairs. Others bought more expensive kinds of trolleys and re-designated them as 'mobile beds'. Public services rapidly became a huge industry dedicated primarily to making the output numbers seem as if they were rising. This was achieved sometimes despite the job they were supposed to do, and often instead of it.

The effect of targets is shocking when you look closely at one small corner of public services, like the business of treating patients within four hours of arriving at A&E. This was a laudable objective, of course, but a visit to a busy casualty department soon revealed reveal a series of 'trackers' or 'bed managers' – often reassigned from nursing – to push the patients faster through the system, plus a whole range of staff dedicated to each target at management level in the various tiers of health management. There is then pressure also from the hospital managers to play the system, encouraging doctors to do anything – even taking blood from a patient – so that they can take them out of the statistics to say that treatment has begun. There is also evidence that managers encourage doctors to treat patients who will attract the most revenue.

There has been a slow dawning of a realization about what this obsessive measurement has done to our services, perhaps most of all in child protection – where every new scandal creates more systems and procedures that get even further in the way. 'They brought targets and indicators into a relationship-based service,' said Professor Eileen Munro at the London School of Economics. 'Once they realized the targets were having an adverse effect, they put in other targets to try to counteract it. So it went on until we've reached the point now where professionals do things to keep government happy and are not focused on how to keep a child safe and happy.'

The man who became most associated with government targets in the UK had been organizing the schools' literacy hour. Michael Barber took up his new position as head of the new Prime Minister's Delivery Unit in 2001, squeezed between the Department of Health in Whitehall and the Red Lion pub. Barber soon invented himself as the geek at the heart of the government machine. 'No one in Whitehall was as obsessed as I was with minor shifts on graphs,' he wrote later. He told his daughters that he was the Prime Minister's graphs-drawer. Not for nothing did the columnist Simon Jenkins describe him as a 'mole-like figure'.

Barber called it 'deliverology', and it meant bringing internal political pressure on senior civil servants from each department, highlighting their failures to reach their target figures by using traffic lights – red, orange, green – and reporting back to Blair with the comforting news he craved. By 2003, it all seemed to be working. All the graphs were going in the right direction, except for street crime, congestion and rail reliability (no coincidence that these are all more difficult for frontline staff to manipulate). Barber wrote a book about the experience, *Instruction to Deliver*, and was knighted for his efforts. He now lectures widely on the science of 'deliverology', but his next day job was – you've guessed it – with McKinsey.

But journalists had a growing sense that the figures they were given were illusory. Barber himself describes his presentation in a warm room in the Cabinet Office in 2004. When each graph came up, one of the tabloid journalists at the back whispered to himself 'bullshit'. *Daily Mail* sketch writer Quentin Letts described it as 'comparable to a lecture by the speaking clock'. Ironically, Barber was an advocate of the crucial importance of relationships in making things work, but closer examination of his book reveals that these are not actually relationships with the professionals on the ground, but with the senior civil servants in Whitehall who he was pressurizing. Certainly, his relationships with the commentators could have been better. By the time he stepped down, he had become a reluctant symbol of the government's command-and-control methods.

'Like witch doctors chanting spells, they must recite from efficiency reviews to sustain the electorate's illusion that government is in control of public services which, in reality, are too big and complicated for the most brilliant bureaucrats in the world to manage from a command post in London,' wrote the *Observer* columnist Nick Cohen. He was complaining about the people he saw as the new elite, the number-crunchers, auditors and authors of specifications and targets.

The tragedy of deliverology was that Barber seemed not to have heard of the power of Goodhart's Law. The graphs he sweated over were illusory. Millions of tiny shifts in definition and procedure by every minor civil servant, NHS worker or policeman were making the figures meaningless. It is hardly surprising that the graphs seemed to be going in the right direction. That is what happens when you put intense pressure on the permanent secretary and it filters down the system.

When an anonymous doctor calling himself Nick Edwards published a book in 2007 called *In Stitches*, setting out exactly how the figures were manipulated, the government denied it. When a police officer calling himself David Copperfield did the same for the police – explaining why he was so exasperated by the waste of time and money that he was leaving to join the Canadian police – junior minister Tony McNulty said it was 'more fictional than Dickens'. The irony was that it was precisely this ability to manipulate the figures that led managers to tighten the rules and make the system more complicated still. It was the same in the private sector. Every time staff were caught 'cheating' the system – or every time there was some kind of screw-up – management measuring systems tightened a little bit further.

The real problem was assuming that every public service was simply a version of an assembly line and could be tackled by standardizing responses and turning the people involved into regulated machines. So every time the system tightened up, the chances those brilliant human beings had to make things happen were that much more constrained. It has been a tragic tale of reduced effectiveness bought in the name of efficiency – and it isn't over yet, even under a coalition dedicated to removing targets. Numbers still have a mystical power over officials, who crave their hard-nosed objectivity, forgetting that they are chained to descriptions – words which can be endlessly manipulated. But there are hopeful signs, and one of them is the fact that key people began to emerge, even in government, determined to do it differently.

Louise Casey was one of the most colourful, but also the most successful, of Tony Blair's government appointments. 'They must have been very brave or very stupid,' she said later about her new job, swept into government in charge of tackling homelessness in 1999, from being deputy director of the charity Shelter. She went on to high-profile jobs tackling anti-social behaviour, 'respect' and criminal justice. She was never less than ebullient and controversial. But the most controversial moment came in 2005 when she was secretly recorded at a private dinner uttering the following unministerial phrase: 'If No 10 says bloody "evidence-based policy" to me once more I'll deck them.' The speech was leaked to the *Daily Mail*. It was a wonderfully unlikely thing for a senior civil servant to say, but what did it mean? What, after all, is wrong with evidence-based policy? It must be better than policy based on hunch. Or is it?

Yet there did seem to be a misunderstanding in government about what actually constitutes evidence. If you listened to the whisperings among policy-makers, you found they were on the horns of a dilemma about it. On one hand you heard them complaining that policy was being pushed through without a scrap of evidence to support it. One study found – for all the talk of evidence-based policy – that what really influenced ministers was a human anecdote told to them at a critical moment. On the other horn was the tyranny of evidence, a hard-headed business where sceptical men – it often was men – waited endlessly for the proof that never arrived. As one contemporary said of the philosopher Bertrand Russell, they had an open mind for so long that they couldn't get the damn thing shut.

Which is why, when you looked a little further, to see what evidence actually lies beneath the most hard-headed government departments, you sometimes found very little. Demanding evidence that could never be forthcoming in the way they expected simply meant delaying action. It was a kind of caricature of the McKinsey catchphrase about everything being measurable, the result of managers putting so much faith in the target figures that nothing else seemed real. It was also frustrating, and you could see why Louise Casey was irritated.

She was one of those exceptional outsiders who gets drawn into government and who gets a reputation for making things happen in the phenomenally complex world of modern administration. Her Rough Sleepers Unit was controversial but, by the end of 2002, it had achieved the 70 per cent cut in sleeping rough across England that Tony Blair had demanded (though not quite in London) and was therefore one of the handful of government targets that were genuinely met.

Why did she succeed when so many other objectives failed? Partly because she had the clear support of the Prime Minister. Partly because her controversial pleas not to encourage people to sleep rough by giving them money sparked a genuine debate and partly because it was far better than previous initiatives in getting homeless people to stay in their new tenancies. Those are the conventional reasons, but insiders also pointed to her idiosyncratic style. One colleague described the way she could bring energy and passion to bear on people. 'She is able to influence people to a degree I haven't seen elsewhere, by applying charm and menace in just the right combination,' they said. 'It is an instinct, a natural reaction.'

In short, she was a super-catalyst at the heart of government, and that brought her into conflict with those – inside and outside – who were suspicious about being too effective for various reasons of their own, either because they felt that being a civil servant meant defending the status quo or because they felt it was somehow dishonest to actually *solve* a problem. There were those in the voluntary sector who earned their money out of homelessness, after all, and had a financial stake in the problem continuing. There were also people who felt that admitting progress was being made was somehow compromising the cause. And then again, there did need to be a debate about whether her priorities were right.

But one of the bravest stands she made was to resist targets imposed from outside. She was afraid that they would encourage those at the sharp end to tackle the easiest cases first, to make the figures look good, as they have in so many other areas. For Casey, the key to big changes was to tackle the most *difficult* first. Her objective was to go for those who had lived out of doors for the longest, were the sickest, most drug-addicted and most difficult. If they could get permanent housing for the most vulnerable people, then those who were less vulnerable would be swept up at the same time.

It isn't quite true that the Rough Sleepers Initiative avoided set objectives. They had the clear target of cutting rough sleeping by two thirds and to end the business of putting children in bed and breakfast accommodation. The statistics were also useful ways of shaming local authorities into action: it was possible to use their record to put a mirror up to them. The Rough Sleepers Unit did use that much of Michael Barber's 'deliverology'. But tackling the worst cases first was a key element of their success. So was Louise Casey's ability to be charming and courageous at the same time, which are human skills wielded together in a fearsome way. She may not have distributed chocolate coins, but these are skills along the same lines as Debbie Morrison's. Both led from the front, both made their presence felt by a phenomenal ability to inspire people face-to-face. Casey was present, going out onto the streets herself, making referrals to hostels, shaking up the system in person wherever she felt it needed to be.

If you throw out the rulebook, how do you prevent your staff from making hideous mistakes or simply lazing around? The answer is in building *more* relationships, not fewer. As a manager worried about what is actually happening at the sharp end of the business, you must know that the hard numbers will be delusory. You have to be there, and this is where Louise Casey showed the way forward. John Timpson spends two to three days a week as chairman visiting his branches, getting round all of them every 18 months. His son James, the managing director, does the same. It is also a method used by Jack Welch at General Electric (GE). 'I firmly believe my job is to walk around with a can of water in one hand and a can of fertilizer in the other to make things flourish,' said Welch, who famously never used a personal computer.

This is what Debbie Morrison did in her school, and it is what the management guru Tom Peters recommended under the slogan Management By Walking Around, a technique that he saw working originally at the computer giant Hewlett-Packard. It is the precise opposite of managing by the numbers, or virtually, and it puts relationships with staff at the heart of any organization. You have to be there, asking questions, sniffing the atmosphere, finding out what is actually happening at the frontline, and preferably by yourself. That is what allows people to put their human abilities into effect.

There is a room at the Department for Education where you can stand surrounded by the data that pours in from schools around the country, and feel yourself at the controls of a giant education machine. Nowhere in government is the fantasy of control as strong as it is here, as civil servants and ministers can sit there and feel they are looking at the dashboard of UK education. It is only when you get into real classrooms that the depth of this fantasy is obvious. A friend of mine was a teaching assistant whose teacher, also an Ofsted inspector, spent much of the week in the stationery cupboard because she could no longer cope with the class. Yet the school was moving up the league tables.

This is the kind of delusion that overtook the American military in the first Gulf War, after the first President Bush claimed that their Patriot missiles had achieved a 97 per cent hit rate against incoming Scuds. In fact, it transpired that there was no evidence of any hits at all. Most British doctors tried their new online appointment booking system at least once, though most of them found it didn't work and never tried it again, but officials still managed to delude themselves that 98 per cent were using it.

This is a strange version of group-think. It insulates those in its grip, blinding them to the possibilities of the people they employ. The truth is that Frederick Winslow Taylor and Henry Ford invented a system that, when it is applied to services, only *seems* to increase efficiency. The huge compliance industry, that helps companies measure their way to compliance with international legislation, fosters a peculiar obsession with figures at the expense of what is actually happening. The senior associate dean of the Yale School of Management, Jeffrey Sonnenfeld, has complained that compliance companies such as Institutional Shareholder Services and GovernanceMetrics International, which test for compliance with good board governance regulations, would pass some of the most troubled companies of the past ten years, but mark down the most innovative, like Southwest Airlines and eBay. It causes a kind of corporate self-obsession, a disdain for staff and customers alike, those recalcitrant human elements can seem intent on subverting the measurements.

Along with the fantasy that they can see everything that is happening in the smooth, humming machine they have created, the figures feed another fantasy that managers can see where the complicated causes and effects are in the system. This is the fantasy that they can predict what is going to happen. But they have reckoned without the ideas of Samir Rihani.

Rihani is an experienced planner, NHS director and university lecturer. He was born in Baghdad, but has worked most of his life in public services in and around Liverpool. He first went there in 1967, to work in planning and development, at the height of the technocratic belief in predictive tools. But his intellectual journey from there to his current work, as an international expert on complex systems and public services, began as technical secretary to the a grand project called the Merseyside Area Land Use Transport Study when he was director of transport planning for the region in the late 1970s. The study launched him on a journey that made him doubt the whole technocratic business of prediction – and to another reason for junking the targets culture.

The study took on a Canadian company to predict the population of Merseyside in 1991, and – although Liverpool's population has famously been dropping for the past half century – they fed various formulae into their computer, and the answer was that it would rise by between

200,000–400,000 people. The various committees behind the study were delighted. More population meant more infrastructure projects, more government money and more prestige. But Rihani had doubts, and he was quite right – it didn't happen.

'I had to ask the company what the print-outs meant and how they reached that conclusion, but nobody could understand it,' he says. 'When I said it didn't make any sense, I was absolutely hated. There were many eminent people there, so I had doubts about what I was saying myself. So, being young and practical, I shut up.'

The experience led him to doubt the illusion of order and control, as well as the academic computer models seemingly able to provide a precise picture of the quarter of a century ahead, while the old neighbourhoods came tumbling down and the tower blocks shot up. Meanwhile, the poor public who were supposed to be benefitting from this were already taking millions of decisions that meant the future would be very different. He came to see planners as what he called 'stargazers'.

'I worked with the stargazers for many years,' says Rihani. 'But by inclination – and possibly because of a childhood spent in the Middle East, where most things were left to the capricious will of God – I suspected that life was moulded just as much by chance as by science.'

He began to apply complexity theory to public services and development, and realized that change happens, not because the experts or managers plan it, and not because it is in the strategic plan, but because of millions of human interactions between individuals. Tiny improvements can happen, but only if the conditions are right over a long period of time. The conditions are that people need the freedom and space to make these human interactions happen. To deal with the complexity of making things happen, people need simple regimes that allow them to get on with it – whatever it happens to be.

The point is that people acting as catalysts, who want to make things happen, need a combination of freedom and opportunity if they are going to do so. They need to be able to act as they think they need to – and especially when they are dealing with other people – even if it is teaching children to be life coaches like Debbie Morrison, or cutting keys for them like John Timpson's staff. To make relationships happen, people need simplicity.

It was this idea that drove the man behind the retail chain Marks & Spencer for most of the 20th century, Simon Marks. In 1956, he visited one store at night and found two girls bringing the stock cards up to date, and was appalled. The cards were abolished, along with time clocks. Good relationships were the basis of effectiveness, he said: 'To depend on statistics is to asphyxiate the dynamic spirit of the business.' His campaign was called Operation Simplification and it underpinned the energy of M&S as it entered the era when one in four children's socks in Britain were being bought there. Simple management and simple hierarchies can make all the difference between success and failure for the super-catalysts.

A similar revolution happened at Continental Airlines in 1995 when they were close to collapse. Gordon Bethune, the new president and CEO, symbolically burned the rulebook in the headquarters car park while the staff watched. It was a way to persuade them that they were allowed to think for themselves. Chucking out the rulebook, as Rule 2 demands, is after all just a means to an end – to bring back those human skills that can make things happen. It is a means to creating a sense of common purpose and responsibility at work, whether it is in the public or private sectors.

The evidence increasingly suggests that it is this sense of belonging and community at work that really makes employees creative. That was the discovery that Harvard professor Teresa Amabile made, and she described it those essential ingredients as 'joy and love', which really only a powerful network of relationships can provide. In fact, the whole idea of management is being questioned – a 19th century idea that enforces compliance but has so far struggled to provide the kind of creative relationships people need to make organizations really effective. Until they can get there, despite all the rhetoric about employee empowerment, organizations are still wasting their workforce. A study by the American consultants Towers Perrin in 2008 found that 71 per cent of the employees who responded felt disengaged or disenchanted at work. 'Companies aren't just wasting people's skill and knowledge,' wrote the business pioneer Jeffrey Hollender. 'Their stultifying cultures actively discourage people from contributing more, even though most employees say they are keen to do so.'

The debate about junking targets in government services is at a different stage in the USA, partly because they went through the targets era earlier than we have in the UK, and also because they faced their own budget crisis earlier, during the Clinton years – and it became horribly obvious to them that too much control was actually massively wasteful.

The story goes back even further, to Ted Gaebler, the city manager of the town of Visalia in California when American local government suddenly found its budgets slashed disastrously, and something had to change. In 1978, the state voters backed a law called Proposition 13, which cut property taxes in half. This meant that, along with everywhere else in California, the town's local taxes dropped overnight by a quarter. It led to a crisis in local government, not just in California, but right across the USA.

Gaebler emerged as a spokesman for a new kind of local government, which was more flexible and entrepreneurial. 'Civilized society cannot function effectively without effective government – something that is all too rare today,' he wrote. 'We believe that industrial era governments, with their large, centralized bureaucracies and standardized "one size fits all" services, are not up to the challenges of a rapidly changing information society and knowledge based economy.'

He was writing at least a decade before that kind of rhetoric became standard on both sides of the Atlantic, and long before those standardized services had been set in concrete by the procedures we live with today. For Gaebler, what he really needed was to remove layers of bureaucracy so that his managers could move quickly and use their initiative. He organized a new budget system to let them do that. If they needed to, they could move amounts inside the budget from one line to another. They could also keep what they didn't spend from one year to the next, to end that flurry of pointless spending at the end of the financial year.

So, after the 1984 Los Angeles Olympics, when his parks staff heard that an Olympic-sized aluminium training pool was being sold, he and the assistant superintendent of the school district flew straight down to look at it. When it was clear they could get it half-price, but that two colleges were also after it, they sent off the cheque straight away for the deposit of $60,000. The school district had saved the amount from previous years. Flexible budgets are still an alien concept in most businesses, but they were crucial to the entrepreneurial movement in the USA.

Gaebler was a protégé of James Rouse, the imaginative American developer who invented the 'festival marketplace' concept. His watchword was: 'How can we profit by solving this problem?' This attitude fed into the movement of energetic social enterprise in the USA, which has only fitfully made it across the Atlantic. It also emerged as a book, *Reinventing Government*, which Gaebler wrote with the political consultant and writer David Osborne. Osborne went on to inspire, and then write, the final report of the National Performance Review, which Al Gore turned into a weapon of reform while he was vice-president. Among those who met Gore in 1993 and persuaded him to back the idea of a review to re-invent government was Bob Stone.

Stone had been the Pentagon's deputy assistant secretary for defence for installations, working out that about a third of the entire defence budget was wasted because of bad regulations, probably amounting to $100 billion a year. He experimented by cutting the regulation book for forces housing from 800 pages to 40. One commander asked permission to let craftsmen decide for themselves which spray paint cans could be thrown away, rather than having each one certified by the base chemist. It was Stone who wrote the original principles that would dominate the National Performance Review.

Stone's experience at the Pentagon coincided with the revelations of the cost of simple items when it went through armed forces bureaucracy. The $7,622 coffee percolator bought by the air force was the most spectacular, but the one that really caught the public imagination was the $436 hammer bought for the navy, or – as the Pentagon called it – a 'unidirectional impact generator'. One of the first schemes the Review launched was an annual Hammer Award for public sector employees who had made huge efforts to work more effectively.

The Performance Review spread the word on junking the rulebook by telling horror stories. One of the most famous of these was about the Occupational Safety and Health Administration (OSHA) office in Maine, the equivalent of the UK's Health and Safety Executive. It consistently came top of the league for how much they were doing, for the most punitive citations and fines given out, yet the workplace safety in the state was the worst anywhere. It began to dawn on managers in Maine that there might perhaps be a connection between his failure and the tight control they were exercising. Could it possibly be that all those exhausting audits and inspections and detailed rules were actually making matters worse?

When they had been convinced, they created a small revolution. They set aside the rulebook, and – like Louise Casey – began by tackling the most difficult factories first. They created employee teams in each of them to tackle the safety problems. If the companies agreed to support these teams, they would suspend their inspections and punishments. The result was that the accident injury rate went down by two thirds.

The role of the National Performance Review was to tell stories like that and their regular newsletters were packed with suggestions. Abandon sign-in sheets and clocking-in machines. Buy equipment locally if you think you can get a good price. Waive the need for travel expense receipts for sums under $75. The Federal Reports Elimination and Sunset Act 1995 ended hundreds of reporting requirements, and ended the rest after five years unless they were specifically renewed. The 10,000-page *Federal Personnel Manual* was junked. And, most important of all, public organizations were allowed to recruit people however they wanted.

The Review carried on, in different forms, for the rest of the decade, and many of its ideas crossed the Atlantic, especially the passion for IT. But the British government only swallowed aspects of it. They did not grasp that part and parcel of the whole bundle was the idea of empowering staff by letting go of tight control over them. They did not borrow – as the National Performance Review borrowed – Jack Welch's Work-Out programme for GE employees, designed to train staff to solve problems. Nor did they grasp the importance of tackling the overwhelming inspection regime. But most of all, Whitehall never grasped the idea that, once you had recruited the most brilliant staff you could find, that it might make sense to use their skills to the full.

The National Performance Review had a huge influence on the way government works in the USA. There are still, of course, corners of the most bone-headed bureaucracy and departments dedicated to the most detailed control. If you are the captain of a US Navy aircraft carrier, you can expect the Pentagon to control every move you make. There are pompous, risk-averse officials in every town and city. But American local government is often more flexible, more entrepreneurial, and more imaginative than its UK counterparts, often because it has to be – the budgets are lower and the challenges can be more intractable and much more enormous.

Meanwhile, the British got stuck with the New Public Management, the rigid control by numerical targets and standards, and by the end of the New Labour years in 2010, it had become a hideous waste of resources. We know that the various auditing bodies cost the UK taxpayer about £600 million in 2002, and it must have been far more eight years later. We also know, thanks to accountants PricewaterhouseCoopers, that each local council spent an average of £1.8 million just preparing for an inspection and on showing that they comply with targets, the cost of the effort of collecting figures and reporting back. We don't know the equivalent costs for health authorities, primary care trusts and police authorities, foundation trusts and other local quangos. But if you add that to the cost of the auditors themselves, you might get to a figure somewhere between £4–5 billion a year to pay for the basic infrastructure of target compliance. Nor does that include all the staff dedicated to each target at local level, and the time they take other staff to work with them.

The American reform writer, David Osborne, a trenchant critic of command-and-control, estimated that 20 per cent of American government spending is devoted to controlling the other 80 per cent, via armies of auditors and inspectors. When vice-president Al Gore led the National Performance Review in 1993, they found that one in three federal employees were there to oversee, control, audit or investigate the other two. If you take some estimates that ten per cent of public spending goes on auditing, then it might come to around £50 billion in the UK. There is some confirmation of this because, if you work it out according to Osborne's formula, it comes to somewhere around the same figure. The wage bill for one in five of UK public sector staff is around £48 billion.

The arrival of a new coalition government in 2010 offered some hope that the targets regime would go, and the abolition of the Audit Commission suggests that it has. But it is not clear yet that the new government has grasped the enormous challenge and what it means in practice, because it takes guts – especially for a government – to give up the illusion of control. 'A financial analyst once asked me if I was afraid of losing control of the organization,' said Herb Kelleher, co-founder of one of the very few successful air transporters, Southwest Airlines:

> I told him I've never had control and never wanted it. If you create an environment where the people truly participate, you don't need control. They know what needs to be done and they do it. And the more that people will devote themselves to your cause on a voluntary basis, a willing basis, the fewer hierarchies and control mechanisms you need. We're not looking for blind obedience. We're looking for people who on their own initiative want to be doing because they consider it to be a worthy objective.

That doesn't mean abandoning control altogether. It means shifting from specifying every possible response to setting out broad principles. Some companies use the phrase 'guiding principles', rather than laying down precisely what has to be done. 'Values-based self-governing cultures are inspired by mission and steered by values,' wrote the management consultant Dov Seidman in his book *How*. 'They enshrine long-term principles in place of short-term thinking, and challenge each decision maker to fulfil those principles in every act they perform.' The American retailer Nordstrom provides instructions to its new employees, which includes the sentence: 'Our only rule: use good judgement in all situations.'

Dumping the tight control can't just mean taking no notice. Quite the reverse, it means setting a framework within which people can work, and being there to help them do so. The key issue is what you do with super-catalysts once you have managed to attract them, train them and develop them. The German sociologist Max Weber talked about bureaucracy as an 'iron cage', and if you put super-catalysts in one of these, then you waste them. They have to be able to meet people – either customers or other staff members – and deal with them day-to-day and preferably face-to-face, if they are going to use their considerable skills to make things happen.

Of course there must be safeguards and certainly there has to be oversight – proper oversight on the front line – but otherwise, for goodness sake let them get on with the job.

Find out more

Dov Seidman's book *How* (John Wiley, Hoboken, 2007) is the key text here. The example of the Bristol Drugs Centre comes from Sophia Parker and Joe Heapy, *The Journey to the Interface* (Demos, London, 2007). The remarks from John Timpson are from my interview with him in 2009, unless they are from *Riding a Giraffe* (see Rule 1). I also carried out an interview with Louise Casey and her colleagues that same year. Samir Rihani's book *Complex Systems Theory and Development Practice* (Zed Books, London, 2002) is a good introduction to how complexity applies to public services. He also has a website called www.globalcomplexity.org.

I have written quite widely about the business of targets and measurement and how they miss the point. My book *The Tyranny of Numbers* has some of this (HarperCollins, London, 2001) and there are different examples, including a chapter on Taylor, in the American edition (*The Sum of Our Discontent*, Texere, New York, 2001). My website has a section on this which brings the argument more up to date (www.davidboyle.co.uk/systems). I should also say that the best description of Goodhart's Law ('Any observed statistical regularity will tend to collapse once pressure is placed upon it for control purposes') is in Keith Hoskin's 'The awful idea of accountability' (in Rolland Munro and Jan Mourtisen (eds) *Accountability: Power, Ethos and the Technologies of Managing*, Thomson, London, 1996).

A description of Professor Amabile's research on people's moods at work is in 'The power of ordinary practices' (Michael Roberts in *Working Knowledge*, Harvard Business School, 20 September 2006). The key text on deliverology is Michael Barber's *Instruction to Deliver: Fighting to Transform Britain's Public Services* (Methuen, London, 2008). Dr Nick Edwards' book is called *In Stitches* (The Friday Project, London, 2007). The David Copperfield blog appeared as a book *(Wasting Police Time*, Monday Books, London, 2006)

My sources for the costs of the compliance system are also Michael Power's pamphlet *The Audit Explosion* (Demos, London, 1994) and

Simon Jenkins' pamphlet *Big Bang Localism* (Policy Exchange, London, 2004), which attributes the £600 million figure to Dan Corry at IPPR (Institute for Public Policy Research). The figure of £1.8 million on reporting for each local authority came from the Pricewaterhouse-Coopers report *Mapping the Local Government Performance Reporting Landscape* (DCLG, London, July 2006).

Ted Gaebler's story is told in *Reinventing Government* (see above) and I recommend David Osborne's sequel *The Price of Government* (Basic Books, New York, 2004). More on the National Performance Review from Al Gore's report of it (*Creating a Government that Works Better and Costs Less*, Diane Publishing, Washington, DC, 1993). See also the National Partnership for Reinventing Government that succeeded it (www.fedgate.org/fg_npr.htm).

Finally, Jeffrey Hollender's book, from where the quotation comes, is *The Responsibility Revolution: How the Next Generation of Business Will Win* (Jossey-Bass, New York 2010). See also www.jeffreyhollender.com.

Rule 3

Put relationships at the heart of organizations

A human being with no daemon was like someone without a face, or with their ribs laid open and their heart torn out: something unnatural and uncanny that belonged to the world of night-ghasts, not the waking world of sense.

(Philip Pullman, *Northern Lights*, 1995)

Leaders don't create followers, they create more leaders.

(Tom Peters, 2001)

Summary

- If efficiency means taking out the relationships, then *the more efficient an organization becomes, then the less effective it is.*
- People might be good or bad, but either way, you get worse results – and dangerously worse results – if you control them too closely.
- 'Processes are important, but processes without spirit are fundamentally useless.'

Dr Jenkins was a Cardiff GP whose name has gone into academic literature as the origin of 'Dr Jenkins' hunch'. This was a story that came to the attention of the healthcare professor and general practitioner Dr Trisha Greenhalgh, who added some details of her own, trying to explain how doctors actually work out what is happening to their patients. Dr Jenkins wasn't his real name, but the story was basically true.

It began with Jenkins in his surgery, taking a call from a mother of a family he knew quite well, saying that her little girl had diarrhoea and was behaving strangely. Nearly all children with diarrhoea have a virus and there was no conventional evidence that this was anything else. Yet he was worried enough to break off from his surgery and visit the family straight away, and then to get the girl into hospital. There it was clear that she actually had meningococcal meningitis, and it was caught early enough to save her life.

Dr Jenkins had good reason to be pleased with himself for this astonishing feat of diagnosis, especially as most GPs see just one case of meningococcal meningitis for every 96,000 consultations or so – and Dr Jenkins hadn't even seen her. Maybe it was just the word 'strangely' that alerted him. Maybe the family never usually complained about anything but, even so, strange behaviour is pretty common among children, heaven knows. So how did he work it out, when any computer system looking at the facts would have reassured him enough to deal with the child later?

Dr Greenhalgh borrowed the story to explain what she called 'narrative based medicine', the idea that there was more to diagnosis than just analysing a few facts on a computer, or using any formal guidelines or rules. The stories that patients tell about themselves have a wealth of detail and contextual information that often goes way beyond the basic facts. Just like an x-ray can't tell you very much without the story that goes with it, so the facts need a bit of elucidation.

Narrative-based medicine is suddenly fashionable among academics. They collect the things patients say verbatim and set anthropologists onto them. They also see it as an antidote to the official 'evidence-based medicine', which lies behind the IT and management control systems that now oversee the business of doctoring. Of course, hunches should be based on evidence, but – as in the business of government – the official mind defines evidence increasingly narrowly. In health, it is usually just based on the World Health Organization's classification of approved diseases and their likely statistical outcomes. Medicine is actually a good deal more complicated than that: patients bring a whole series of interlocking symptoms, lodged in a series of stories, which doctors have to unravel. That's the human reality.

But Dr Jenkins and his hunch have a different importance for this book. You can dismiss the story as intuition, which is another of those vital human skills that can get removed by the wrong kind of management systems, but it makes far more sense understood as the result of a long-term relationship with a family. Jenkins knew the family and knew what was normal language for them. He knew their normal tone. His intuition was based on detailed knowledge, but so detailed that it could never be written down on any notes or computer data entry. It may not have been the result of a face-to-face encounter, but it was the result of a face-to-face relationship.

You don't just need people with strong human skills to make organizations work. Nor is it enough to remove the rules and processes and let them get on with the job. You need to let them build relationships with people too. This is important in so many areas of work – from selling to teaching – but it is especially so in medicine. Yet the way we train doctors to record things can corrode all this.

This is the frontline of a debate inside the medical profession, which is probably more intense than it is anywhere else, though it is relevant to any organization. On one side, doctors are trained to standardize medical histories, and define medical problems so they can be diagnosed efficiently, treated and also priced. On the other side are the medical practitioners who value the slightly untidy relationships they have with patients, who are uncategorizable individuals, after all. They know, for example, that only 10 per cent of patients with hypertension have it in its 'standard form'. They know that only half of what GPs see can be classified according to standard biomedical taxonomy. They know that medical students get trained to boil down the stories they hear until they eventually stop noticing the bits that can't be recorded formally.

Patients are human beings. They are never quite 'standard'. They always break out of the categories of management systems. Of course, there are many aspects of medicine where a simple transaction is all that is needed, but – for anything beyond that – a doctor-patient relationship without a relationship is going to be less effective, more time-consuming and more expensive. Doctors need to be able to build relationships with patients if they want to understand or influence them.

We have seen that the whole idea that human relationships lie at the heart of making things work is anathema to conventional wisdom. Like human beings, relationships reek of disorder. They can be uncontrollable. They complicate definitions, which is the way that modern medicine is managed – especially in the USA – defining diseases by standardized symptoms, using huge population databases. In the USA, relationships also complicate the usual business of pricing. Doctors defer to a brief conversation with the insurer, just long enough to slot the symptoms they were told about into a payment plan, which makes relationships with patients largely irrelevant.

'Severing the classic doctor-patient relationship is Job One under a system of covert rationing,' wrote an American doctor, blogging under the name Dr Rich. 'Doctors simply cannot be allowed any longer to place their patients first. They've got to place the needs of their true masters first.'

But the real front line of the battle is in schools. The irony is that, like in medicine, our education system is full of caring, imaginative people who exercise their skills in building relationships every day. But there is no doubt that the same misplaced understanding of efficiency that regards human relationships as a complicating factor is as powerful here as it is in medicine. In fact, the radical educationalist Michael Fielding describes the educational equivalent of evidence-based medicine under a heading of 'Naming the new totalitarianism'. This 'high performance' schooling regards any personal relationship as important only in so far as it helps achieve the specific target numbers or as far as it enables teachers to tick the boxes on 'learning outcomes' in any situation. Relationships between pupils and teachers are then subsumed into the technocratic business of generating numbers, of chopping the business of education up into tiny fragments that can be tracked and ticked.

On the one side is the machinery of high-performance schooling; on the other side is the human business of educating individual pupils, their individual needs and aspirations, and their relationship with teachers which achieve this. Fielding warns that the worst thing is that, increasingly, it is hard for those at the heart of the system to tell the two apart. In the end, the rationalized targets systems drive out any other knowledge or understanding:

> Our practice of schooling where the optimism, energy and goodwill of contemporary approaches are leading us down a road that, albeit unintentionally, is likely to produce a society that diminishes our humanity, destroys much that is of worth, and denies much we seem to desire.

That is one way to explain how we have found ourselves in a world where teachers are encouraged to get children to count adjectives or metaphors rather than telling them stories. It threatens to produce an education system, according to the children's laureate Michael Morpurgo, that is 'focused on the mechanics of literacy not the experience and the joy of literature'. In teaching, as in medicine, the whole idea of stories seems to rather scare those in the grip of illusory efficiencies, maybe because it reminds them of the possibilities of human complexity. Maybe it reminds them, as Philip Pullman might say, of the days before their *daemon* was removed. Then they lose their faith in human beings and they panic in the face of what are emphatically human problems that are not really amenable to technology.

We already have seven-year-olds who are rarely allowed outside the classroom or into the natural world, or encouraged to paint or make music, but are lectured for an hour about how to put glitter onto glue (I've seen them). Or who have to draw from miserable photocopies rather than from life, or who have bits of stories for comprehension rather than the whole things. Yet the vast majority of teachers remain committed to pupils as individuals, and they resist the idea of reducing education to modules delivered by computer, or by interactive screen, to largely passive and uncritical children. In fact, there is something of a backlash against the sheer dullness of education delivered to children in this way, chained to their desks, doing nothing practical, but provided with information, information, information, like geese being fattened up for *foie gras*. The battle continues.

This is the meaning of Philip Pullman's analogy of *daemons*. It is the question of what lies at the heart of human endeavour, whether there is anything there worth discussing, or whether chopping up human processes into their constituent functions and deliverables will destroy that heart forever, at great financial and human cost. And behind all that is the question of relationships, and whether they are relevant to effectiveness.

The evidence that relationships *are* important is all around us: intractable customer service lines, failing schools, bored children, chronic ill-health, repeat offending, and all the other frustrations of modern organization. We have all seen them, but we tend to dismiss each of them as one of those irritating aspects of modernity, a kind of by-product of progress. The truth is that, if efficiency means taking out the relationships, then *the more efficient an organization becomes, then the less effective it is*.

Doctor-patient relationships go back to the days of Hippocrates, who even told doctors to taste the body secretions of their patients if it helped them make a diagnosis, not something you see much these days. In most of our mega-hospitals, we rarely see the same doctor twice, let alone find them sipping our body fluids. In the USA, the shift by managed care insurers away from personal doctors is a huge political issue.

There certainly are those, often young people, who prefer the convenience of just being able to drop in and deal with whatever isolated problem they have. Others feel very strongly that they want a continuing relationship with one doctor. One of the leading UK advocates of doctor-patient relationships points out that the national average with our GPs is eight and a half minutes, five times a year.. This actually amounts to more than nine hours with one doctor (if you stick at one doctor) over a decade or so. You can really get to know someone in nine hours.

But the managed care regime in the USA, which has been a model for some of the recent developments in the UK, has been particularly corrosive of this relationship, because insurance companies – the ubiquitous health maintenance organizations (HMOs) – can decide which doctor you will see. They have also driven down the length of consultations to seven and a half minutes, where the doctor interrupts on average after 23 seconds. A generation ago, busy doctors used to see maybe 30 patients a day. In the new managed care system in the USA, it is not unusual for them to see 70.

There can certainly be a downside for doctors and patients if the relationship is too intense or exclusive. It can be frustrating for the doctor if they feel they are just maintaining a patient in ill-health. It can get in the way of getting an objective second opinion. On the other hand, the evidence is mounting that some kind of continuing relationship with a doctor can make people get better quicker. Research shows that it means patients stick better to treatment, go to hospital less and stay there for less time. It means doctors tend to be more committed to making the bureaucracy work for individuals.

We know that a loving relationship between a parent and a baby actually turns on the neurons in its brain. We know that people who have no close relationships are risking their health as much as heavy smokers. All the evidence is that relationships keep you healthy. The relationship between teachers and pupils does at least some of the educating, and in the same way it is your relationship with your doctor that seems to do some of the healing. That certainly seems to be the result of a strange experiment carried out in a GP's surgery in Devon.

Dr Michael Dixon is now prominent as the chair of the GPs' lobby group the NHS Alliance. Ten years ago, as now, he ran a clinic in the town of Cullompton in Devon. The seeds of the experiment came when he went along to a course in complementary medicine, mainly because he felt guilty catching himself throwing their leaflet away. 'I was very unpersuaded by the language,' he says. 'But I was persuaded by the slightly different way they viewed patients.'

It was an important discovery for him, and the beginning of a path that would see him embracing acupuncture and various other ideas in his practice. But it also happened to coincide with the visit of a local healer, the wife of a judge, who arrived at the surgery and asked for a job. Dixon was fascinated, and especially with the possibilities for research. The thing about healers is that they don't use pills or needles. It is just them, their hands, and their relationship with the patient that counts. It was also clear quite quickly that Jill the healer was making a difference.

'She clearly was having an effect on people who had been ill for a long time,' he says. 'Quite apart from anything else, it seemed to be changing their attitudes. They were suddenly prepared to go off and try things when before they had been fatalistic. Jill was empathizing with them, giving them an ego boost and making them feel important. She was doing the things that good doctors do, but it is much more difficult doing that in five or ten minutes than in half-hour sessions.'

The anecdotal evidence was crying out for some kind of research, but did he dare – when doctors are routinely castigated from a puritanical minority for dealing in what they call 'mumbo-jumbo'? But what the research would be looking at was not so much whether healing works, but whether the sessions with the healer were effective. It would be a research study, as much as anything else, on how important a doctor-patient relationship is. So, on a more systematic basis, he began offering his patients the chance of ten healing sessions. After the weekly sessions on Thursday mornings, the research team were amazed to find that as many as 80 per cent of the patients felt better. More than half said they were 'much better' and stayed much better the next time they were asked three months later, even though – looked at under a microscope – the problems still seemed to be there.

Quite apart from the question of whether the healing worked, the clear implication was that taking regular time with patients doesn't just matter a bit, it matters a very great deal. 'One important job as a GP seems to me to be to change people's perspective,' says Dixon. 'Healing for people seemed to help them turn the corner in their attitude to their illness.'

The research was published in the *Journal of the Royal Society of Medicine*, and sceptical rage poured down on the heads of Dr Dixon and his team, but as a study of relationships it is one of the most dramatic. 'It is vital,' he says now. 'If someone is overweight, really the only thing that will change them is a relationship. If you can get under their skin, and they can get under yours, then you can almost lead them on their own terms – but that is all about relationships. It doesn't work just to give them a diet sheet.'

The trouble is that, in the UK as in so many other places, those who pay doctors on behalf of the public have less interest in this aspect of their work. 'I am paid for my quality formulae scores, for my access figures and a whole range of other things,' says Mike Dixon. 'But nobody really pays me for building an effective relationship with you, or listening to what you say. All that is taken for granted.'

A lot of the controversy about this is based on a misunderstanding among doctors about what patients actually want from them. A survey by *Which?* magazine showed that most doctors believed patients wanted top expertise and a variety of services, when actually most of them wanted someone to give them some time – which is what the healer did. 'The patient deserves to be known as a human being,' says Professor Bernard Lown from the Harvard School of Public Health, 'not merely as the outer wrappings for a disease.'

I am sure Lown is right, but the evidence is that people's health will also get better if they are known as such. Which means that a 2009 survey of patients at the University of Chicago Hospital is particularly worrying. As many as three quarters couldn't name any doctor who had treated them and, of the other quarter, 40 per cent got the name wrong.

We have known since 1970 that a good relationship with your doctor makes for more successful chemotherapy. We have known since 1968 that informal human relationships make for more effective decision-making in business. We have known for decades that the most efficient manufacturing plants are those with fewer than 45 staff, where they can really know each other. But, for some reason, the vital importance of face-to-face relationships at work has been quietly forgotten, and the result of removing the human is that human endeavour becomes that much more difficult.

It also corrodes what trust there is between customers and staff, and that can be hugely expensive. A report by the Health Service Ombudsman in 2007 said that that the number of formal complaints was so high in the NHS (138,000) because managers were often unable to behave in a human way. They said 'a bit of courage and common sense' could have resolved most of the complaints before they reached that stage. Given that the NHS paid out £800 million in 2008 in damages (not including legal costs), that has huge cost implications in itself.

One hospital that has tackled this is the University of Michigan Hospital in Ann Arbor, which faced a crisis in 2001. In that year, they were facing more than 260 lawsuits, at a possible cost of $8 million. There wasn't much they could do about accidents that caused death or real injury, they agreed, but minor mistakes really shouldn't go to court. Instead, they encouraged doctors to say 'I'm sorry' and to do so on the spot.

Their lawyers warned the hospital that this was the equivalent of legal suicide, but they were wrong. Over the next three years, the hospital's lawsuits dropped by half, and the cost of the ones that continued also dropped by half. Actually, people could forgive mistakes if they were made by human beings, but when the system closed up and denied everything, it absolutely enraged people. Rebuilding those ordinary human relationships saved a great deal of money.

The extreme end of the spectrum of control is the CIA, which is so terrified of relationships – or to be more precise, of infiltration – that spies have to fill in pages of forms, signed by three different levels of managers, whenever they meet a foreigner. Background checks take more than a year before they can start their jobs – and they are often rejected if they have relatives abroad or have travelled anywhere too interesting. The ideal candidate is a Mormon with little foreign experience. After 9/11, it transpired that the desk officer in the Pentagon responsible for key parts of the Middle East did not speak Arabic. They relied instead on technology or databases instead of relationships.

'The inevitable, if unintended, consequence of the government's reliance on computer databases,' wrote Jill Kirby of the Centre for Policy Studies, 'is that the tracking of information is replacing personal interaction between professionals in which the sharing of information is mediated by human contact. Such contact – the conversation between two head teachers about a member of staff moving from one school to another; the discussion between social worker and doctor about a child's injuries, or the taking up of references on volunteers – is crowded out or in some cases actively discouraged by systems requiring the impersonal exchange of data.'

That is the point. It may be management, targets or software. It may be a fond belief in a particular kind of efficiency among IT or management consultants. It may be such an overweening pride in the exclusivity of the profession that they don't like building relationships with clients, either because they prefer to have an intermediary between them (barristers) or because they believe their creativity might suffer (some architects spring to mind). Whatever it is, it frustrates progress if it gets in the way of long-term face-to-face relationships. Only then can the super-catalysts do their work.

When the management theorist Naresh Khatri came to live in the USA, having been brought up in India and Singapore, one thing particularly intrigued him about the American health system: why did everyone talk about how much it was changing when, as far as he could see, it wasn't changing at all?

Khatri had been in the USA before as a student, and – to him at least, and despite all the rhetoric and cacophony of change – nothing actually seemed to be any different. Certainly there were changes of regime. There were the new health maintenance organizations and the rationalization of diagnoses and symptoms. But the basic feel of surgeries and doctors was much as it always had been. There were still hugely expensive lawsuits, and the same vast insurance bills, the same huge hospital corporations and the same mistakes. Nearly 100,000 patients died in the USA every year because of mistakes, more than car accidents and breast cancer (in England and Wales, the equivalent figure is contested, but is usually put at around 2,000 a year).

Khatri had no inside experience of healthcare at that stage. He had worked at the Federal Bank of India for five years. But like any good academic researcher, he set out to find out why there was this mismatch between his perception and everyone else's. He began to organize seminars on healthcare among his fellow management academics at the University of Missouri. He tested his ideas against similar research in other countries, and the conclusion he came to was that hierarchical management style, the blame culture, and the obsession with top-down IT systems, had just carried on regardless. If the new systems, which caused so much argument, were really more efficient, you might expect that mistakes would be going down too. Actually they stayed much the same. Health reformers were even using the burgeoning cost of hospital mistakes as a reason for standardizing and controlling even more, but there was little evidence that they were right.

Khatri began to suspect that the highly-controlled culture of compliance and blame was actually why the mistakes were happening in the first place. So he and his team designed an experiment to categorize 16 hospitals in Missouri by management style to see if it had any effect on hospital mistakes. They also put in a whole series of tests to see if they were thinking along the right lines, including surveys of more than 1,000 health providers across the USA. The results were peculiar: the expected link between medical mistakes and a culture of blame wasn't there in the sample. But what was clear was that there were fewer mistakes when the medical staff trusted and felt good about each other, and more drug-related errors in hospital cultures which were exerting the most detailed control.

Highly controlling management cultures (Admiral Tryon-style management) meant more mistakes than looser control (Admiral Nelson-style management). Controlling management meant narrowly defined jobs, time clocks and tick-box training programmes. It assumed that people are not capable of regulating themselves. What Khatri called 'commitment-based management' meant self-regulation based on trust among staff:

> The current bias towards innovative technological solutions over those that require the transformation of current dysfunctional culture, management systems, and work processes in healthcare must be corrected if medical errors and quality of patient care are to be taken seriously.

This confirmed the idea that control is less effective than letting staff use their human skills, though it doesn't explain why this might be. But other research that was going on at the same time suggested a reason. Another study found that up to 80 per cent of hospital mistakes in American hospitals had less to do with technical problems than with the personal interactions inside the healthcare teams. Working in a blame culture forces staff to protect themselves, even if it is just against reams of paperwork. They put more effort into shifting blame than genuinely discussing mistakes, or what are called in the jargon 'adverse events'. In other words, it is their relationships with each other – and with their managers of course – that make the difference. This isn't about face-to-face relationships with patients, it is about face-to-face *working* relationships.

When mistakes happen in a controlling culture, where relationships are curtailed, the result is a kind of vicious circle. There has to be more control, more system and more blame, which means even less motivated, more disaffected staff, and more mistakes, and so it goes round. On the other hand, when the control is relaxed and replaced with more imaginative leadership, people can start going beyond their remit. When the relationships work, they take more responsibility, show more initiative, morale goes up and the staff turnover goes down.

'We tend to think that, without regulation, people will do stupid things,' says Khatri. 'But actually, they don't. And when you exercise that kind of control, then you are not using people's ideas fully.'

Management theory has traditionally fallen into two camps, known loosely as Theory X and Theory Y. One is based on the idea that people are basically bad, the other that they are basically good. What Khatri's research suggests is that this is the wrong question. People might be good or bad, but either way, you get worse results – and dangerously worse results – if you control them too closely. At the very least, you lose half the abilities of your staff. But if you replace that control with some kind of encouragement for self-control, based on trusting relationships, then you not only cut down mistakes, you also engage people's imagination and creativity too.

Here we get to the nitty-gritty of the argument. The People Principle suggests that human beings are critical to making systems work. That is the meaning of Debbie Morrison and her chocolate coins, of Captain Foley at the Battle of the Nile, and of the other stories too. They all understand that human beings and the relationship between them are the critical missing factor in modern projects. It isn't that human ingenuity and imagination are all-powerful instruments, but they do trump organizational systems. Understanding this marks the beginning of the end for putting process before people.

We have come to the end of a period of management history where systems were put at the heart of organizations, public and private. The next era of organizations still seems undecided about it, but there is a growing sense out of that that – when the systems did their job effectively – there was usually a passionate individual at the heart of it. Tom Peters, the management writer who first set off in a more humane direction, argued that it is qualities beyond the right process or the right formula that really make things work. Talking about his business heroes back in 2000, including Mike Walsh who obliterated the bureaucracy at Union Pacific, he said:

> Mike was a process fanatic. Herb Kelleher [Southwest Airlines, see Rule 2] is a process fanatic. Welch is a process fanatic. But they have the totally leavening influence of pure passion, albeit the Kelleher passion is expressed in a different way than Welch's. But if you scratch Herb Kelleher or Jack Welch, they bleed real blood, and Mike Walsh was the same way. And the companies that get screwed up are the one who brought in [corporate consultants] ... and didn't have the passion. Processes are important, but processes without spirit are fundamentally useless.

This is an important distinction, and one which is tough for big organizations simply because you can't turn it into a formula. It is also a summary of the case for Rule 3. There are certainly forces at work that have been pushing the world in the other direction. But the success of companies such as Timpson and others are demonstrating a new way forward. Nobody will take any notice of them if the systems most organizations use now are working well, but they're not and – as Ralph Waldo Emerson put it – if you make a more efficient mousetrap, the world will beat a path to your door. There has never been a moment in world history when we have needed human skills and ingenuity as much as we do now.

There are obvious objections to Rule 3. Trust employees too much and they will mess things up, sit around drinking, steal from the till and so on. Rely too much on individuals and they will misuse their position, cause legal problems or worse. All that is true. These things will happen sometimes. But the evidence is mounting up that – if organizations can employ the right kind of people (not necessarily the best qualified) – then it doesn't happen very much. More than that, by putting face-to-face relationships at the heart of what systems survive, then organizations usually avoid these pitfalls.

On the other hand, we can't simply go back to the processes of a generation ago. Targets and systems may have excluded people's imagination, but they were also brought in to tackle the sheer unfairness of the systems we lived with when I was growing up – and those dark, outrageous corners of discrimination and hopelessness that existed then. We accepted then that surgeons could experiment without restraint, without questioning them. We accepted those hideous psychiatric wards where old people with Alzheimer's were simply sent to moulder away. All that systematization was partly the business of refusing to settle for second-best. It succeeded in making those abuses obvious; it just hasn't made the leap forward we need.

Find out more

The story of Dr Jenkins' hunch is in *Narrative-based Medicine: Dialogue and Discourse in Clinical Practice* (Trisha Greenhalgh and Brian Hurwitz (eds), BMJ Books, London, 1998). Also, more recently, *Narrative-based Healthcare: Sharing Stories: A Multi-professional Workbook* (Trisha Greenhalgh and Anna Collard, BMJ Books, London, 2003). The original research which suggested that only half of what doctors see fit into standard diagnoses is in K. B. Thomas 'The temporary dependent patient' (*British Medical Journal*, vol I, pp59–65, 1974). There is more anecdote about how over-defining diagnoses is damaging patients on an entertaining blog by an American doctor calling himself Dr Rich (http://covertrationingblog.com).

Michael Fielding's latest thinking is set out in his book with Peter Moss (*Radical Education and the Common School*, Routledge, London). His article about personalization and the philosophy of John Macmurray is hugely important in this area and deserves a wider readership ('The human cost and intellectual poverty of high performance schooling: Radical philosophy, John Macmurray and the remaking of person-centred education', in *Journal of Education Policy*, vol 22, no 4, July 2007).

As for doctor-patient relationships, the key text is *The Human Effect in Medicine: Theory, Research and Practice* (Michael Dixon and Kieran Sweeney, Radcliffe Medical Press, Abingdon, 2000). There is a summary of evidence in 'Towards a theory of continuity of care' by Denis Pereira Gray and others (*Journal of the Royal Society of Medicine*, April 2003). Dr Dixon's healer's experiment is written up in 'Does healing benefit patients with chronic symptoms? A quasi-randomized trial in general practice' (*Journal of the Royal Society of Medicine*, April 1998). There is increasing literature about the costs of distrust, but the two examples here were taken from Dov Seidman's book *How* (see Rule 2).

There is also increasing literature about the links between control and mistakes. See Naresh Khatri's research article 'The relationship between management philosophy and clinical outcomes' (Naresh Khatri, Jonathon R. B. Halbesleben, Gregory F. Peroski and Wilbert Meyer, *Health Care Management Review* vol 32, no 2, April/June 2007). See also his 'Medical errors and quality of care: From control to commitment' (Naresh Khatri and others, *California Management Review*, vol 48, spring 2006).

Demerge everything

Whenever something is wrong, something is too big.
(Leopold Kohr, economics professor, visionary and former roommate
of Hemingway and Malraux, 1957)

*When efficiency is measured by return on assets, smaller businesses
are more efficient than larger businesses in every industry.*
(US Treasury Department, 1967)

Summary

- If we want our organizations to work more effectively, we have to
  end the tendency of all of them to strive towards empire.
- Big system managers fall into the many illusions that are encouraged
  by their own systems, because big organizations require systems in
  a way that small ones don't.
- It isn't that small is always going to be more effective than big, but
  that – because it allows human beings to work effectively and build
  relationships – it often is.

They say that all roads lead to Rome. That may be so, but I know one
thing for sure: all bus routes end in Crystal Palace, a south London
suburb on a hill. I know this partly because I live there and partly because
of the crowds of dazed bus passengers wandering around near the bus
station, tipped out unexpectedly at the terminus, some miles short of
their homes. But it does mean I get a choice of buses where I live, and
two in particular go nearly past my house. I tend to catch whichever one
comes first, but there's a big difference between them.

If I get the 450, it is a small single-decker bus. It is usually packed full of pensioners hanging onto the supports for dear life as the driver swings them round. When everybody gets off, young and old, they often thank him. If somebody very old is clambering aboard, someone usually leaps off and helps them on. Sometimes the driver does. People also chat to each other, which is very unusual, even in Crystal Palace.

But if the 468 comes first, it is a different experience. It is the same cross-section of passengers, but this is one of those broad red double-decker buses, the motoring equivalent of a brontosaurus. The atmosphere is completely different: nobody talks. Nobody even smiles. Nobody speaks to the driver, though he occasionally swears at us. Nobody helps anyone on or off. You can cut the distrust with a knife, and unfortunately people occasionally do.

I have thought about this a great deal. It isn't that there are more regulations on the smaller bus; the driver ignores them anyway. The government doesn't regulate smaller buses more intensely. It is subject to no extra government targets or funding for social cohesion. We have undergone no extra community training. Yet I can't help feeling that, if we could package whatever quality it was that made the 450 friendly, we might solve the nation's crime and community problems. But I don't suppose many Whitehall ministers travel along the 450 route to Thornton Heath. It has the air of a mandarin-free zone.

No, I think the success of the 450 is more about its size. It is smaller and, because it is smaller, the drivers' human factor comes into play. We are aware of them as people: they can – and often do – make people's day. On the 468, the poor benighted drivers are forced to be adjuncts of their machines. Like the policeman in Flann O'Brien's novel *The Third Policeman*, they are already part metal. 'Size seems to make many organizations slow-thinking, resistant to change, and smug,' said the great investor Warren Buffett in his 2007 letter to shareholders. He was talking about companies, but it applies equally well to buses.

This is not a scientific experiment. The 450 and 468 don't follow exactly the same routes; one route might involve more schoolchildren – but the phenomenon is instantly recognizable. We know from personal experience that there are knock-on human effects when systems get bigger. We have all stood behind an elderly customer at the supermarket checkout trying to chat with the cashier about the weather – just as she used to when the small shop was there – while the queue behind taps its collective feet. We know what happens when organizations are too big: the systems take over.

In fact, this is another critical factor that can make or break organizations that employ people for their human skills. Organizations built on an inhuman scale, with their inhuman architecture and terrifying marble lobbies, don't find it so easy to provide the kind of simple regime that allows human beings to make things happen. That is why, if we want our organizations to work more effectively, we have to end the tendency of all of them to strive towards empire. Rule 4 says we have to split them up.

For the last century or so we have come to believe that the unspeakable problems of society, such as poverty and social collapse, happen in cities. So it comes as some surprise to find a secondary school with very impoverished pupils in the heart of rural England, in this case in the Golden Valley of Herefordshire. But that is what Fairfield High School is, in one of the worst pockets of rural poverty anywhere in the country. One in four of their 267 pupils has special educational needs. Yet Fairfield is also one of the top specialist schools and academies in all three categories of achievement, in raw exam results, value added and overall high performance. The latest letter from the school inspectorate Ofsted is filled with words like 'exceptional' and 'outstanding'.

When Chris Barker became head in 1999, promoted from the English department, he was told during the interview that the job might not last long. The school had been designated for closure. In January 2008, he got a letter of congratulation from the government on the same day as another letter from the council warning him that the school would shut. In both cases it was for the same reason: the school's size.

Let's leave on one side for a moment the economics of all this. Local authority funding arrangements have disadvantaged smaller schools for decades. But professional advice in the case of schools and hospitals is also usually for larger units, because it is one way they can give greater choice and specialist equipment, but also because they mean higher status and salaries. Even so, Fairfield is less than a tenth of the size of many other big secondary schools in the UK, so it gives the lie to the idea that small schools must provide a worse education.

In fact, a great deal of what Chris Barker could provide (he stepped down in September 2010) was simply because the school *is* relatively small, where he could know every pupil by name and provide for what they most needed on an individual basis. There are animals all over the site, and Barker found a BTEC qualification in animal care which could really engage the local children. The pens and fences were built with the help of pupils studying for the construction BTEC. When he bought some extra land to enlarge the site, he was quoted £2000 a year to cut the grass. He bought some alpacas instead.

None of this is in the conventional manual for headteachers. Nor is the boy who loved flowerbeds: Chris Barker promised that, if he wanted to do a BTEC in horticulture, he would come in early to teach him before school, so that he could fit it around his other subjects. But then this kind of rule-breaking is possible in small schools where everyone knows each other. The sense of informality means that people go out of their way to make things happen. But there is no doubt that Barker is also a super-catalyst, but he is also able to use his human skills to inspire because the school is small enough for everyone to know everyone else.

You can see the same factors at work in the well-known way that people look out for each other in small communities when they don't in big ones. In 1993, a television documentary filmed an elderly actress falling down in the street to see how long it would take for someone to help her up. In the village it took two minutes, but in a busy shopping mall it took nearly three quarters of an hour. Small communities help human interactions along, with all the disadvantages of that as well as the advantages – sometimes people want a bit of privacy, after all.

Of course, this sounds a bit glib. You can imagine companies, factories, schools, hospitals or doctor's surgeries that are just *too* small, and rely too much on one individual. We all know communities that are too small, inward-looking or actually in-bred. I certainly do. What we have to do here is to strike a balance so that institutions stay human-scale. That is certainly confirmed by most research into small schools over the past generation, which has challenged the idea that schools are better when they are bigger. Despite this, for the past generation or so, most policy-makers have *believed* that big schools are better. They seem to have started thinking this in the USA after the successful Soviet launch of the Sputnik spacecraft. They persuaded themselves that somehow only huge schools could produce enough scientists to compete with the USSR. It is one of the peculiar ways that Soviet thinking filtered into the West.

The first challenge to it came from Roger Barker, describing himself as an environmental psychologist, who set up a statistical research centre in a small town in Kansas after the Second World War and researched the local schools to within an inch of their lives. It was his 1964 book *Big School, Small School*, with his colleague Paul Gump, which revealed that – despite what you might expect – there were more activities outside the classroom in the smaller schools than there were in the bigger schools. There were more pupils involved in them in the smaller schools, between three and 20 times more in fact. He also found children were more tolerant of each other in small schools.

This was precisely the opposite of what the big school advocates had suggested: big schools were supposed to mean more choice and opportunity. It wasn't so. Nor was this a research anomaly. Most of the research has been carried out in the USA, rather than the UK, but it consistently shows that small schools (300–800 pupils at secondary level) have better results, better behaviour, less truancy and vandalism and better relationships between staff, pupils and parents than in bigger schools. They show better achievement by pupils from ethnic minorities and from very poor families. If you take away the funding anomalies that privilege bigger institutions, they don't cost any more to run.

But why should smaller schools work better? There is some consensus among researchers about this. The answer is that small schools make transformational human relationships possible. Teachers can know pupils and vice versa. 'Those of us who were researchers saw the damage caused by facelessness and namelessness,' said the Brown University educationalist Ted Sizer, who ran a five-year investigation into factory schooling in the 1970s. 'You cannot teach a child well unless you know that child well.'

Frightening evidence of this came in June 2008, when the *Times Educational Supplement* reported that 21 per cent of Year 8 pupils said they had never spoken to a teacher. 'Talk to the children, if you can,' one school volunteer I know was told by the headteacher on their first visit. 'Nobody talks to them these days.' That is evidence for a lot of peculiar things about our society, but it is also about the scale of organizations, and schools in particular.

Sizer is behind the recent decision in Boston's education department to dramatically cut the size of their schools. Of course, it isn't just about size: you can't just have small schools that ape the systems and facelessness of big schools. 'Otherwise small schools are just big schools in drag,' says the principal of Boston Arts Academy, one of two small schools next to the Red Socks stadium in Boston. You can build relationships in big schools too; the point is that it is much more difficult.

Debbie Morrison's second headship (see Rule 1) was at a school nearly three times the size as Mitchell High in Coventry, and the chocolate coin regime became impossible, so she had special thank you cards printed instead. 'Never in 30 years of teaching have I ever been thanked before,' said one teacher.

'Well,' said Debbie, 'do the same for your team.' She now sends birthday cards to all 300 staff.

Even in a larger institution, these details of face-to-face relationships have to be done at one remove. It is hardly surprising that it doesn't always work. Bigger institutions are more difficult for working this magic. There is a clue here about why, despite our innate ability to make human relationships, it sometimes takes a super-catalyst like Debbie Morrison to make an impact. The bigger the organization, the more exceptional people have to be to make things work. 'The human touch is still possible,' she says. 'But it has to be more creative and it depends on more people engaging with it. You need to build other energy creators around you.'

Bryant Park is tucked away in the heart of New York City, next to the New York Public Library, between Forty-Second Street and Fifth Avenue. These days it claims to be the most densely used urban space in the world, packed with sunbathers, readers or walkers, depending on the season. But it wasn't always like this. By the end of the 1970s, the garden, re-designed by parks commissioner Robert Moses in 1934 to cover the library's book stacks, had become a symbol of the city's decline, a no-go area, full of needles, drug dealers and muggers.

The fact that it is now such a civilized place, where you can sit safely under a tree on a seat cushion (provided) and read books and magazines (also provided, as long as you give them back afterwards) is partly down to the stringent conditions set on a major grant to the library in 1980 by the Rockefeller Brothers Fund. They said the park needed to be cleaned up before they would release the money. An informal group of people, including the chairman of *Time* magazine set out to do something about it, formed a Business Improvement District – the legal mechanism that allows businesses to agree to tax themselves for a common objective – and set about looking for someone to run it.

Their choice of the 26-year-old Dan Biederman was inspired, as it turned out, because Biederman is definitely a super-catalyst, and now known around the world for his expertise in turning places around. It was said later that they plucked Biederman from obscurity, which is not strictly true – he was a recent MBA graduate from Harvard Business School – but he certainly had little track record for the job he was taking on. 'Taking on' is also the right expression, because he was going to confront a situation which had floored generations of park managers and city authorities.

For the next eight years, Biederman joined forces with Vartan Gregorian from the library, struggling to persuade the various city agencies to let go, the Parks Council, City Council, Arts Commission, Board of Estimate and many others besides. There were hundreds of meetings before they eventually agreed to let Biederman have his way, but they did. Once he was genuinely in charge, his approach was to concentrate on the details. No doubt somebody like Debbie Morrison might have tried to hire the trouble-makers, but Biederman's skills as a super-catalyst lay elsewhere. He hired his own security staff to confront the behaviour of some of the park users. He tackled the plague of pigeons by recruiting a squadron of hawks. He used his freedom to clean up Bryant Park in his own way.

He was fascinated in the 'broken windows' theory, then just articulated – where very small signs of disorder were found to attract bigger ones. Whether this is strictly true is beside the point: that was the principle by which he tackled the problem of the park. He also consulted the great urbanist William Whyte, who persuaded him that he needed the informality of moveable chairs, so that people could arrange them where they needed to go to have a good time with each other. Then he fought a long battle to stop park keepers collecting them up every night. No, the chairs needed the whiff of informality. His biggest struggles were to persuade his staff not to sweep up the leaves – the leaves were natural, it was the litter that needed picking up – and abandoning the official preference for benches screwed to the ground. But by face-to-face contact with everyone with any influence on the park, he achieved it.

The point about Biederman is not just his awareness of the impact of tiny details, but the fact that he is present, walking around the park with a notebook, jotting down ideas for improvements. 'I call it management by wandering,' he says. This is another example of the feature of super-catalysts known as Management By Walking Around. It provides that all-powerful human element, but it also relies on something else – the sheer simplicity of Biederman's chain of command. If he saw something wrong, he could get it fixed. He didn't have to ask permission, or balance it against a string of other objectives. He could do it. This is a clue about why people are effective using their human skills in some organizations rather than others.

The point about this story is not so much that Biederman is a super-catalyst, capable of making things happen via face-to-face relationships, though he certainly is. It is that he has managed to escape the sheer complexity of bureaucratic systems that constrain so many of those like him around the world. His success came because of the simplicity of his remit, and the simplicity was possible because the park, the system and the organization he controlled were small.

It is an important lesson given that our generation, and the most recent ones gone by, have been absolutely addicted to size, whether it is their homes, bedrooms, cars or fridges – and in particular, we have allowed organizations to become bigger and bigger. Four banks (Citigroup, Chase, Wells Fargo and Bank of America) now issue half the mortgages and two thirds of the credit cards in the USA. Tesco hoovers up nearly a third of all the grocery spending in the UK. Five companies control 90 per cent of the world grain trade. Six control three-quarters of the global pesticides market. The people who run these organizations are, of course, paid more the bigger they get. But the truth is that small-scale scares them. They fall into the many illusions that are encouraged by their own systems, because big organizations require systems in a way that small ones don't.

May 2008 was unseasonably hot. Morning after morning, I sat down in my local railway waiting room only to find the heaters on. On swelteringly hot days, I found myself climbing on the chairs to switch them off – only to find them back on again the next hot morning. I began to wonder why this was. Perhaps the passengers switched them on in the cold early mornings, and there was nobody to switch them off. Perhaps the station staff – who I had never seen on the platforms – had instructions to switch them on.

But I did have a clue. During the hottest week of the month, my mother-in-law arrived at the council-run college in Croydon where she taught part-time, to find that the central heating was on. It was particularly sticky and sweltering. During every spare moment, she set about the long business of tracking down somebody who had sufficient authority to turn off the radiators. My mother-in-law is one of those people who can make things happen, very gently but determinedly, but – even for her – getting the radiators turned off in the sweltering heat was no easy project. The principal of the college wasn't responsible. Nor were those responsible for the college at the local authority. Most of them not only had no power over their own heating, they also had no idea who had – a familiar experience in centralized public services.

Towards the end of the day, she discovered the right person. It was a man with a laptop, somewhere in the council building that also housed the education officers. He was persuaded to act, and – at the click of a mouse – the radiators went off.

In those heady first few years after the Berlin Wall came down, I used to write a regular newsletter on renewable energy, and often included anecdotes about the energy use in the great Soviet-style apartment buildings on the outskirts of Moscow or Budapest – pumping heat into the surrounding atmosphere whether it was hot or cold. We used to laugh at this, amazed that nobody could turn off their ancient totalitarian radiators. Yet we seem to be in a similar situation in the UK – my wife was teaching in a local school where the radiators were also blazing out during the hottest days, so I don't believe this is actually very unusual. The reason is the same; the institutions were too big for the human element to work.

The most obvious thing to say about all this is that it is hugely expensive. What proportion of the public sector heating or lighting bill is unnecessary because nobody feels responsible, or nobody knows the local situation? What does it cost a Canary Wharf bank to keep its lights and computers blazing all night, because nobody feels responsible for turning them off? What are the costs of those mistakes, that we have all encountered, when local authorities paint the windows of a council estate just two weeks before the windows were due to be replaced? It is absolutely impossible to know. But just because these extra costs of bigness are unknowable, it doesn't mean we can discount them.

But the real point is that smaller units are more likely to let humans to use their unique skills effectively, building relationships with each other, summing each other up. It also means they can pay attention to detail, as nobody was doing in the story of the Croydon radiators. That does not necessarily mean a whole plethora of tiny units, but it does mean that breaking down huge units into smaller ones – and giving people effective responsibility over them – tends to be more cost-effective.

It explains the peculiar phenomenon of why public clocks so rarely work, why railway stations are so unkempt, why car parks are so hideous. Because nobody has responsibility for them. It also explains the peculiar phenomenon, first noted two centuries ago by the radical reformer William Cobbett, but confirmed since: why is it that ten farms of 100 acres each produce more than one farm of 1000 acres, and why is the produce more varied? Nobel prize-winning economist Amartya Sen has shown that small family farms are more productive than big industrial ones. Economies of scale? Hardly. The personal touch, attention to detail, the effectiveness of human-scale housekeeping over other economic systems? It isn't quite clear, but the same thing works in other areas of production too.

Big organizations mean different kinds of behaviour, but also different kinds of people. They have more imperial people at the top, who are paid vastly more (how else can Mark Thompson justify his £834,000 salary as director-general of the BBC?). They have more docile people lower down. The founder of Delta Consulting, David Nadler, has written about working for the telecoms giant AT&T when they began looking to recruit more entrepreneurs, only to find that people like that don't tend to work for companies like AT&T.

There is another problem that, the bigger the contracts to deliver services, the more distant are the relationships between service providers and consumers. The managers become even more remote, and the prices rise – how could they not when procurement is being bundled up to be managed by huge private agencies and when there are only a few contractors capable of bidding for contracts of that size? Public sector managers Capita launched the notorious Criminal Records Bureau checking system in 2002, seven months late, and there was almost immediately a backlog of 300,000 people waiting for quite unnecessary criminal record checks, preventing them from taking up posts in schools or running voluntary organizations. But there was no other company capable of organizing a project on that scale. The same increasingly applies to charities in the UK, 50 of which signed a letter in 2008 complaining that only organizations with an income of £400,000 a year were allowed to apply to the new Empowerment Fund. Only 2 per cent of UK charities now hoover up two thirds of the funding.

A 2009 report by Reform found that the cost of policing in the UK had gone up by £4.5 billion since Labour came to power in 1997, when the costs of compliance and diseconomies of scale began to rise. That may be one reason why the smallest police forces are the most effective, catching more criminals for their population than the big ones. That is another reason why American hospitals cost more to run the bigger they get. These are the costs of scale in the public sector.

There is some evidence of the costs of size in the private sector too. When the business writer Robert Waterman says that the key to business success is 'building relationships with customers, suppliers and employees that are exceptionally hard for competitors to duplicate', you know things will have to shift. Because size gets in the way of that. There is evidence that when companies get bigger – and more impersonal – then they are less able to be innovative, which is why so many pharmaceutical companies are outsourcing their research to small research start-ups. In fact, this trend seems to have been going on for most of the 20th century. Half a century ago, the General Electric finance company chairman T. K. Quinn put it like this:

> Not a single distinctively new electric home appliance has ever been created by one of the giant concerns – not the first washing machine, electric range, dryer, iron or ironer, electric lamp, refrigerator, radio, toaster, fan, heating pad, razor, lawn mower, freezer, air conditioner, vacuum cleaner, dishwasher or grill. The record of the giants is one of moving in, buying out, and absorbing after the fact.

It is true that two musicians invented Kodachrome film in a bathroom. Google was invented by two computer nerds in a university student digs. Big organizations are not creative. So why do they stay big? There are occasionally de-mergers of companies. Cadbury and Schweppes went their separate ways after nearly four decades together. Sir John Harvey-Jones famously dismembered ICI. But generally speaking, corporates get bigger, largely because the salaries and share options are more lucrative and the earnings available for people organizing the mergers are out of all proportion to their benefits. The accountants KPMG studied the result of mergers and acquisitions (a mega-industry worth $2.2 trillion) in 1999 and found that shareholders lose out in more than 80 per cent of all cross-border mergers. They found that only 17 per cent of all mergers added value to the combined company, while as many as 53 per cent actually destroyed shareholder value. So why do they happen? Because the rewards of those individuals who make them happen are absolutely huge.

This is the problem. Managers of big corporations are paid inordinately more than managers of small ones. Managers of big schools, hospitals and police forces get paid much more, and have more status than managers of small ones. Consequently, despite all the evidence that they cost more and are less effective, schools are still getting bigger on both sides of the Atlantic. The number of British schools with more than 2000 pupils has tripled in the last decade. As I write, a new school is opening in Nottingham with more than 3500 pupils. Some schools in New York City have a terrifying 5000 in them. That is why UK hospitals are gobbling up each other and why – despite all the evidence – we seem likely to see mergers of police forces. It is a terrifying testament to the power of self-interest over effectiveness.

The justification for huge impersonal organizations is usually that they make 'economies of scale' possible. Ever since Frederick Winslow Taylor and Henry Ford's factories split people's jobs into their constituent parts and ran assembly lines past them, we have had the idea that creating big organizations can be cheaper than small ones. That has been the justification for merging schools or hospitals. One big institution should have fewer costs than a handful of small ones.

You can see why this should seem to be so. If you merge organizations, you can cut out some of the duplication. You can sack some of the duplicate marketing managers. You only need one accounts department. But what you also get are the extra costs of bigness, such as the radiators in Croydon, and all the other inefficiencies of a bigger management infrastructure, more rules, more monitoring, more paperwork, more databases and more IT support. That is why, despite the commitment of Whitehall and the big consultancies, the idea of economies of scale is not working any more. 'Beyond very small volumes [it] is a concept that should be discarded,' said accounting professor Tom Johnson.

Economies of scale works well in assembly lines when parts and products are standardized. That is why it was such a breakthrough for Henry Ford. But even car assembly plants have moved away from the idea. Toyota tries to reduce its batches right down to one at a time. It is the flow through they want, not the huge scale. But service industries, dealing with human beings, can't standardize. When they try to do so, they find so many exceptions – which are so difficult to deal with – that the savings start to evaporate. This is the story of all those shared-back office services and call centres that we now live with.

But there is another problem about bigness, which is that big organizations rely more on bureaucracies – and bureaucracies have an in-built tendency to grow. Bureaucracies tend to serve themselves and the bigger they are, the bigger their self-serving becomes. They are places where managers can insulate themselves most successfully from the real world, where they can really be convinced by those target figures that pour in, and where they are that much less tolerant of people who *can* see the reality around them. They are also more likely to confuse the demands of their executives – for more data or more status – with the needs of their client groups. Their tendency to grow also seems to apply whether they are succeeding or failing.

The historian C. Northcote Parkinson, who wrote his management classic *Parkinson's Law* in 1958 to explain that the work expands to fill the time available, had bureaucracies in his sights when he did so. Parkinson famously looked at the British Admiralty, and discovered that there were 2000 civil servants working there at the outbreak of the First World War in 1914 to administer a navy of 146,000 seamen. By 1928, just 14 years later, that figure had grown to 3569 civil servants to manage 100,000 seamen. There are now only 38,000 personnel in the navy, and they are serviced by only 1790 civil servants. On the other hand, those are just the ones assigned entirely to the navy, so we do have to delve a little deeper. You would also have to assign some of the 19,200 civil servants running the Ministry of Defence in London, and some of the 14,300 civil servants looking after defence equipment. Of course, modern missiles and computers are more complicated than shells and range-finders, but Parkinson's Law still seems to apply. 'The number of officials and the quantity of the work are not related to each other at all,' said Parkinson.

I remember when I applied for a job to edit the quarterly 16-page newspaper for one of our biggest environment lobby groups, I puzzled over this question, trying to work out why this required a full-time post (I was at the time editing a weekly eight-page newspaper, and doing so part-time, with the help of a proofreader and nobody else). Having failed to answer the question by the time we got to the second interview, I made the mistake of asking. 'There doesn't seem to be quite enough to do,' I said, naively discounting the impact of bureaucracy. I didn't get the job.

This is not a small issue. One of the factors blamed for the collapse of the Roman Empire is that the thousand or so administrators and bureaucrats required to run the imperial administration under Augustus had swelled to somewhere around 32,000 four centuries later. Bureaucracies do that. Things get more complicated, whether or not the task is any more complicated than it was, and this growth makes it harder for organizations to be effective. The archaeologist Joseph Tainter, from Utah State University, argued in his 1988 book *The Collapse of Complex Societies* that every layer of organization increases the energy and human effort involved in making society work. Tainter said that there comes a point when societies need all the energy they have just to maintain the complex systems, at which point they are so overstretched that climate crises or barbarians at the gate bring the whole edifice down.

Organizations tend to serve themselves; absolute organizations tend to serve themselves absolutely. They serve the hierarchy, which adds layers of management, each one of which make it more difficult for people to be effective, tougher for the super-catalysts to work their magic and harder to make change happen. It also makes the needs of the internal system far too important, and often turns the outside world – the clients and customers – into an irritating distraction. It is very hard for a five- or seven-tier council Learning Disability Team to break out of their 'processes' to meet the individual needs of clients, compared to the flexibility of a two- or three-tiered organization. It is hard for corporate leviathans such as Tesco or the BBC to shift from their usual grooves.

This is as true for the private sector as it is for the public. That's why Next shortened its three-month customer service waiting list by simply deleting it. That is also why we are seeing the bizarre new phenomenon of mega-corps 'sacking' their own customers if they get too difficult. The bank ING Direct now asks 10,000 people to close their account every month if they are 'high maintenance' (difficult ones). The giant American health provider Aetna got big by sacking their 'unprofitable clients' (ill ones). One of the main reasons for appalling customer service is the size of the organization. That doesn't mean that big organizations have to be inefficient or unpleasant, but it is that much harder for them not to be.

The lesson of all this is not that smallness is always going to be more effective than big, but that – because it allows human beings to work effectively and build relationships – it often is.

Find out more

The most important book in this area is a little dated now, but just as compelling: Kirkpatrick Sale's ground-breaking *Human Scale* (Secker & Warburg, London, 1980). The classic text on scale is Leopold Kohr's *The Breakdown of Nations* (Routledge & Kegan Paul, London, 1986), which has a foreword by Ivan Illich, who is another key influence on this book, and this chapter in particular (see his *Tools for Conviviality*, Harper & Row, New York, 1973). So is Colin Ward in all his many writings, particularly in his columns in *Town & Country Planning* magazine. In fact, the stories about the 450 bus and about the Croydon radiators both come from my own regular column in the same magazine.

The original research I refer to is Roger Barker and Paul Gump's *Big School, Small School* (Stanford University Press, 1964). The best collection of evidence on small schools is Kathleen Cotton's *New Small Learning Communities: Findings from Recent Literature* (Portland, Oregon, Northwest Regional Educational Laboratory, 2001). There is a more recent American summary by the Chicago Public Schools System at http://smallschools.cps.k12.il.us/research.html. The equivalent UK information is available from Human Scale Education (www.hse.org.uk). I also recommend Linda Jones' article 'My daughter's reports are meaningless drivel' (*The Independent*, 26 July 2007).

The quotation about innovation by T. K. Quinn is in *Nation* (3 July 1953). The figures on the failure of mergers are in a KPMG report (*World Class Transactions, Insights into Creating Shareholder Value Through Mergers and Acquisitions*, KPMG Transaction Services, London, 2001). Finally, Joseph Tainter's book is *The Collapse of Complex Societies* (Cambridge University Press, Cambridge, 1986). I already mentioned *Parkinson's Law* in the Introduction.

As I was writing this, I came across an excellent article which covers similar territory, which I thoroughly recommend: Christopher Ketcham, 'The curse of bigness', *Orion*, Mar–Apr 2010.

Rule 5

Obliterate the hierarchies and empires

We are constantly amazed by how much people will do when they are not told what to do by management.
(Jack Welch, former CEO of General Electric)

Most companies suck the life out of people.
(Tom Freston, former CEO of Viacom)

Summary

- Giving responsibility back to staff has an equal and opposite effect on the hierarchy, which then has to transform themselves into something different – no more detailed decision-making and much more mentoring and inspiration.
- 'I have come to believe that economies of scale is one of the most overrated concepts in business. It exists, of course, but it is overtaken by diseconomies of scale much sooner than most people realize.'
- Academics have been studying teams for well over a decade. The problem is that most organizations haven't acted on it. Those that have acted on it have left the control mechanisms and hierarchies largely intact.

Durham in North Carolina is a growing city, of jazz, tobacco and civil rights activism. But there on the outskirts of the town is a General Electric (GE) plant making aero engines, each one with more than 10,000 different parts and manufactured to exact specifications. There are 200 people who work there, most of them in teams of less than 20 technicians. The teams are self-governing. Different members take the lead for different projects. Their only instructions are the delivery date the engines are required. Outside the teams, most decisions are taken by plant committees. There are no time clocks, no cleaners (they clean up after themselves) and no lockable boxes for machine tools. There is very little hierarchy, only three pay grades and no middle managers.

Yet between 1995 and 2000, they cut the cost of making engines by half. They cut defects by three quarters. Only one in four engines now has a single flaw, which is usually cosmetic, such as a scratch mark on the outer casing. The rest are almost perfect. They also release ten pounds of toxic chemicals into the local environment every year, when the equivalent GE plant at Evendale, outside Cincinnati, releases 2000 pounds. 'In Durham, you've got people who think,' Evendale general manager Bob McEwan told *Fast Company* magazine. 'I think what they've discovered in Durham is the value of the human being.'

The lesson of Durham is that there are some heights of achievement which can be achieved as a group, even though they can't always be achieved by people on their own. People trust each other, and spark off each other, but they also make it possible for each other to be creative. Because, in organizations at least, being creative requires there to be appreciative creative people around you. That is where teams come in.

Research by Boris Groysberg and Linda-Eling Lee published by *Harvard Business Review* in 2007 found that, despite appearances (and bonuses), the big investment stars of Wall Street were highly dependent on having very successful colleagues working at different levels of the company. Teams matter, because they act as a kind of adrenalin for human creativity inside organizations. They can have ideas and put them into effect. In fact, one way of making organizations feel smaller is to re-organize into teams that can manage themselves and take on whole jobs.

That is the point, because sometimes it just isn't possible to take an organization apart and split it into small, independent units. But you can get some of the benefits of human scale, and that means splitting the tasks of the organization up into a series of small, multi-disciplinary teams. This brings us back to the maverick 20-year reign of Jack Welch over General Electric, the presiding genius over the GE Durham plant. He knew he could never make GE a small organization, but he knew he could make it smaller and also that it could *feel* smaller. 'Small companies waste less,' he said:

> They spend less time in endless reviews and approvals and politics and paper drills. They have fewer people; therefore they only do important things. Their people are free to direct their energy and attention toward the marketplace rather than fighting bureaucracy.

Of course, GE was no small company. Getting to feel like a small company was therefore a huge undertaking. But Welch believed it was the way to set people free to be catalysts. 'The talents of our people are greatly underestimated and their skills under-utilized,' he said. The point was to end the architecture of control, understanding that it was actually delusory.

'None of us runs the business,' said Welch. 'I'm never going to run them. I don't run them at all. If I tried to run them, I'd go crazy. I can smell when someone running [a business] isn't doing it right.'

Note that business of 'smelling' whether things are right or not. In a lesser manager, this might just look like bluster but, in someone of Welch's record, it is more like intuition. You need that if you are Managing By Walking Around. That is how you build relationships with the people who work for you. It can be misused, of course, but it can't possibly be replaced by processes. Only being on the spot, and knowing people, can give you that sense. And it helps if the organization is small, or – at the very least – is informal enough to *feel* small.

As I said in Rule 2, none of this prevented GE from serious environmental abuses under Welch. Nor did it lead to the kind of obliteration of hierarchies that proper team working needs. But it did happen in their Durham plant, and that has become the main inspiration for this radical new way of organizing. What is peculiar is that, despite the publicity about the success of GE Durham, so few organizations have risked going the same way – especially in the public sector where quality and attention to detail can really be a matter of life and death.

Despite their damaging addiction to mergers, mainstream business did latch onto the idea that headquarters could be massively shrunk. It saves money on office space and it can potentially cut bureaucracy too. One of the first companies that really acted on this was the Swedish power giant Asea Brown Boveri in 1988, when chief executive Percy Barnevik slashed headquarters staff from 2000 to 175. He set what he called the '90 Per Cent Rule', which meant that 90 per cent of the staff in headquarters, and regional headquarters too, must be re-assigned to the front line or to spin off. The same applied to the research budget, 90 per cent of which was to be decided without waiting for permission from frontline managers.

Few other companies have managed quite such a ruthless downsizing, but companies such as IBM and BP have famously made huge reductions to their headquarters staff, and it remains a controversial thing to do. IBM now sends 40 per cent of their headquarters staff out of the office. BP removed a whole tier of management so that line managers now report direct to the board. Union Pacific famously shed most of their layers of management. Pfizer managed to cut their layers of management from 14 to eight.

Some researchers suggest that downsizing has gone too far, so that crucial management functions are allowed to rot – or are done virtually, which undermines the ability of executives to meet staff face-to-face. But the downsizing trend has still had little real impact on the size of corporate entities. Giants such as Walmart and Tesco are still tolerated, despite the negative impact they have on local economies (Walmart gets more than $2 billion in subsidies a year from the US government in income support to their employees). Goldman Sachs got a $650 million subsidy to rebuild their corporate headquarters in New York. We may have smaller headquarters, but we still have the same giants living in them, and most of them have entirely ignored the lessons about scale.

But there are exceptions among the biggest companies the world, especially among those – such as Apple – which found that developing breakthrough products was very much more effective when it was done in small teams. The most famous example is WL Gore.

WL Gore began as a company in 1958 with one patent, for a new kind of cable, when it was launched by a former DuPont executive called Bill Gore and his wife Genevieve. It is now best known for its Gore-Tex anoraks and other innovative uses of plastics – this was a company that really took the advice to Dustin Hoffman in the film *The Graduate* to heart ('One word: Plastics') – but also for the innovative way it has divided its employees up into teams.

This began in 1967, when Gore designed a 'lattice' structure for the company, which he regarded as the antidote to the hierarchy he had endured at DuPont. He had noticed how well the *ad hoc* teams had worked developing new projects or solving problems, and he worked on shaping the idea for the next decade. Since then, there have been no conventional chains of command at WL Gore. Nor are staff given bosses. Instead, they choose to follow specific leaders of teams.

Like Timpson, there are no heavy manuals of rules. Instead, there is a general duty on staff to communicate with each other, to keep their commitments to each other and to consult with their colleagues more widely if they want to do anything that might undermine the company, its profits or its reputation. It is an unusual, fluid structure for an organization that employs more than 8000 people at 50 sites around the world, but it seems to be the basis for the company's huge success in recruiting and keeping staff. WL Gore has been voted one of the 100 best places in the world to work in *Fortune* magazine for the last 12 years.

Gore noticed that, whenever there was a crisis, most companies set up a task force to tackle it and the first thing they did was throw out the rulebook. It seems to work for crises, so he wondered why he needed to wait for one. The result is a system where there are no job titles. There is almost no hierarchy. The Gore culture demands that anyone in the company should be able to speak to anyone else. It is a federation of teams rather than a monolithic empire, and the idea is that this informal atmosphere – created by these interlocking small teams – is the secret of the company's innovative record. Staff are encouraged to chase ideas on their own, if they want, and get together to develop them.

Leaders are not appointed as such. People become leaders by leading. Nobody has to follow you – you have to use your abilities as a catalyst to attract people into your project and make things happen. This is what Gore calls 'natural leadership'. Pay is decided by committees according to the contribution they think people make.

The result is that the company was able to plunge into the music business with a new kind of string for acoustic guitars, an idea which emerged out of their medical products when they applied it to improving the gear cables on mountain bikes. The company puts its technologists and its salespeople in the same building in their headquarters complex in Delaware, in the hope that this kind of synergy would happen. The experiments to find a better guitar string carried on for two years without much success, but they gathered a small team around the idea and, a year later, the product was ready to take onto the next stage. The team never asked permission to do it. They had no need to ask superiors to watch their experiments or to account for them. But the result was the best-selling Elixir strings.

There is no doubt that working in teams can be disconcerting and even occasionally nerve-wracking. Some management writers talk about them as being as authoritarian as hierarchies. Gore insiders talk about spending their first six months in any new team just trying to prove themselves to their colleagues. It is true that team working is not necessarily any more gentle than hierarchies, but it does seem to work and play to people's strengths.

WL Gore's experience of team-working may not have had much impact on the structure of organizations, but it has launched a huge academic literature about what size the teams ought to be. At one end, there seems to be some consensus about the size of the organization or department, which ought to hover around 150 people. Psychologists agree that 150 is about the limit where everyone can know everyone else. When Anita Roddick watched The Body Shop grow, it was at that point that she said she no longer knew the names of her employee's pets, and the company began to grow into a more impersonal organism.

At the other end, most studies agree on the number five, partly because five is the limit of human short-term memory. People can hold teams of five in their heads at any one time. The truth is that it all depends on the task that needs to be done, and every team is going to be different. What is most important about team-working is not so much that it is effective (though it is), or that you can somehow calculate precisely the right number (which you can't), it is that team-working uses people's human skills to the full – especially, according to some studies, when they have close friends on the same team. It is no coincidence that the proposed optimum team size is related to human psychology, because human-scale organizations work better. 'In the battle between company policy and human nature, human nature always wins,' wrote Rodd Wagner and James Harter in their business bestseller *12*. 'Companies do far better to harness this kind of social capital than to fight against it.'

Behind Rule 5 is one of the practical problems about putting people back at the heart of organizations. If you recognize that human beings using their personalities to brilliant effect is the key to making organizations work properly – that super-catalysts are the solution – then that does pose a problem for existing managers. If you recruit the right employees and let them get on with it, in the way they know best, then what should the managers do? It isn't surprising that there is going to be some resistance from them. What on earth is their role? How do they justify their salaries?

This question keeps coming up as we work through the implications of people power. The answer seems to be that giving responsibility back to staff has an equal and opposite effect on the hierarchy, which then has to transform itself into something different – no more detailed decision-making and much more mentoring and inspiration. This is a different kind of leadership, providing the catalyst for change, and breaking out of the illusion that you are somehow pulling the strings to make the details happen yourself.

WL Gore is not the only company to obliterate their hierarchies and build teams instead. Linden Lab, the creators of the virtual world Second Life, works on a similar basis. Employees are expected to choose a task, tell everyone else what they are doing at the start of the week and how they got on at the end. Everyone reads the reports. Clothing company Patagonia lets employees organize their time themselves. So why, in those circumstances, are we still building vast public service organizations, with intricate and disempowering hierarchies, which sap the will and initiative of the people who work in them?

Ricardo Semler is a Brazilian businessman, though he is actually half Austrian, who took over his father's company at the age of 21. This was not a comfortable experience for either of them. He was frustrated with the traditional hierarchy of the company, Semco, and determined that it should diversify out of the shipbuilding industry (they made pumps) before it was too late. When he told his father he had to leave, Antonio Semler was deeply upset. He decided to give his son a majority share of the company and to go on a prolonged holiday. 'Whatever changes you want to make in the organization,' he told him, 'do them now.'

What happened next is one of the strangest stories in business, leading to a whole new model, which obliterates all the divisions and hierarchies, and with dramatic success. It is one of those ubiquitous case studies that are learned, almost by rote, in Harvard Business School, but which has been hardly copied anywhere.

The young Semler responded to his father's challenge. He summoned up all his courage and sacked most of the company's executives, including his father's closest friends, and set about changing Semco, eventually turning it upside down and creating a whole new model of doing management, what is in effect a democratic company. The company not only survived the experience – though it did so by the skin of its teeth – but it also became one of the most successful companies in Brazil, making white goods and a range of other related products and services. From revenues of $4 million in 1982, when the young Semler took over, it now earns more than $200 million a year, with an annual growth rate of up to 40 per cent. It employs 3000 workers, and has managed to sustain that growth through Brazil's years of hyper-inflation and its years of disastrous deflation as well.

There were a whole series of false starts, but the real turning point came when Semler suffered a series of fainting fits when he was 25 and was told by his doctor that he had to simplify his life. It coincided with the sense of frustration that he was feeling about the conventional ways of controlling companies. 'Semco appeared highly organized and well-disciplined,' he said. 'But we still could not get our people to perform as we wanted, or be happy with their jobs. There weren't enough cathedral builders.'

He decided that, not only was his own work-life balance completely flawed, but also that his employees were working far too hard as well. A semi-autonomous unit he set up inside the company to develop new products and business ideas was so successful that it provided a template for re-organizing the whole company. By the 1990s, Semco had emerged with a revolutionary new structure, with semi-independent units and factories run largely democratically, where the employees work in teams without job titles. They choose the hours they work, and even choose their own salaries.

This has taken Samir Rihani's idea of simplicity to a level that has barely been tried anywhere else, but it seems to work for Semco. One of the keys to this was to be as open and honest as possible. It meant that everyone had to know what everyone else earned, and there had to be no secrets about the company's financial position either, because all employees have a stake in it. Of course, choosing your own hours of work can have potential pitfalls, especially if you are in a factory and you need to organize regular shifts. But Semler says that, in practice, people sort it out between them. 'We set up a task farce to mediate any problems,' he said. 'It hasn't met yet.'

Semco organizes regular courses for employees in accounting just so that they can understand all the financial information they are given. For very big decisions, such as moving factories or buying companies, the whole workforce votes. When they move a factory, they hire buses and everyone goes and inspects the sites. All their internal meetings are voluntary, and anyone can come if they have got something to say. If they get boring, people can go. In fact, if nearly everyone has left by the time they come to the question of what people should do as a result, they ask themselves if the project was really worth doing in the first place.

All this is anathema to most business lobby groups in the world – almost as much as letting staff choose their own hours and salaries. Other companies around the world have followed in the same direction of complete openness, including regular open evaluation of managers by the people they manage, but the democratic element is very unusual. In small, family firms, including my own (I am the only employee) then the same flexibility and democracy applies. In big companies, it is almost unique, though there are mutually-owned companies such as John Lewis and Scott Bader where the staff own the organization.

Semler jokes about his own shrinking role and his own shrinking office. After ten years of taking no decisions at all, he held a reception to celebrate the occasion. This understates the struggle he has undergone to transform Semco into a model of people power, against the rigorous opposition of the Brazilian business community, the sophisticated business media, but also – less predictably – the unions, who were deeply suspicious of the idea of people choosing their own working hours.

What Semco has is an empowering sense of flexibility. The rulebooks have been abolished, along with the strategy documents that sit unread in the drawers of most organizations. The manuals have been replaced by a simple rule about using common sense. Like Timpson, many forms and manuals have been replaced by simple cartoons. Semler has described how terrifying it was for people in the accounts department to manage people's expenses without rules, just using their common sense, but they managed. Unfortunately, our own MPs failed to manage the same feat.

'All these rules cause employees to forget that a company needs to be creative and adaptive to survive,' he said. 'Rules slow it down.' Not only that, they divert attention away from objectives, create work for middle managers, and provide a false sense of security for managers. Semco isn't necessarily a comfortable place to work, where you take leadership responsibility if you feel you want to, where everyone can see what you earn and how you perform, and where there is no safe bureaucracy to hide inside – but it is certainly effective. They have an employee turnover rate of only 1 per cent a year.

Semler is characteristically tough about those who cling to the old dispensation:

> If you want my advice, take a deep breath, pluck up your courage, and feed the policy manual to the shredder, one page at a time. Let companies be ruled by wisdom that varies from factory to factory and worker to worker. To do otherwise only gives those tough guy controllers the comfortable feeling that the company is organized and provides jobs for dozens of disturbed souls who should be retrained for some useful purpose.

Openness is important. The UK Parliament has been trying to conjure up a whole set of rules to govern MPs' expenses, forgetting that the real problem was not that there were no rules, it was that the expenses claims were secret. Openness is a challenging business. It means that leaders emerge on their own merits and people must prove themselves worthy of their peers, and must continue to do so day by day. When you internalize the rules, they are much simpler, but in a sense they are also a good deal stricter. The difference is that they are flexible and human and they don't constrain people in Weber's iron cage of bureaucracy. They don't suck the life out of organizations and make them hugely expensive and inefficient to run.

There are other parallels with Timpson's Upside Down Management. There is the same preoccupation with recruiting people for their personality rather than their qualifications, still less their sartorial elegance.

There is the same emphasis on management by walking around, and on face-to-face communication. There is a similar resistance to the idea that they are trying to create an unrealistic paternalist sect. 'We are not a big happy family', Semler says, 'we're trying to be an efficient business.'

The difference between Semler and Timpson is largely what they both emphasize. If Timpson talks about replacing process with people, Semler wants to emphasize getting rid of the illusion of control. It isn't that somehow he has surrendered the ability to control his own company – though clearly he has in some sense of the word – it is that he has abandoned the delusion that he could control what people did there. That is why he urges managers to get rid of their strategy documents: "If I had ten to 15 minutes, I could write a business plan for you, with extrapolation added to wishful thinking,' he says. The point is to confront the powerlessness of most managers head on. 'We don't know what we are doing in Semco,' says Semler, 'but we would rather not pretend that we do.'

But the key to Semco's success is their intricate networks of self-organized teams. Once the teams were in charge of the separate units, the company's systematic organizational chart had to go out of the window. People took on the roles of leaders and convenors because they had natural authority to do so, and to achieve different tasks. Like the other reforms Semler introduced, it required far greater transparency. 'The only way to change is to make each business unit small enough, so that people can understand and contribute accordingly,' he said.

But the other crucial factor was size. To discuss properly, in such a way that everyone could bring what skills they had to bear, the units had to be small enough for everyone to know each other's first names and to feel they belonged.

Semler's new sub-divided factories began to innovate faster. They also managed to deliver products the day after orders were made, which had never happened before. They bounced back from problems quicker. 'I have come to believe that economies of scale is one of the most overrated concepts in business,' wrote Semler. 'It exists, of course, but it is overtaken by diseconomies of scale much sooner than most people realize.'

There are those who believe that something alchemical happens when teams work closely together, and that they can achieve things, and have ideas or solve problems, that simply can't be done by individuals or traditional hierarchies. The doctrine that groups have a kind of intelligence of their own is one of the French contributions to modern philosophy. Much of the thinking about 'collective intelligence' comes out of France, but it is Americans who tend to put it into practice. Wikipedia is updated and refined by tens of thousands of users around the world. Linux and the other open source software projects are developed by virtual teams who have never met. They happen outside any chain of command, building on virtual communities held together by trust.

One of the leading enthusiasts for 'collective intelligence', a former stuntman and co-founder of AOL France, Jean-François Noubel, believes that it goes further than that. That people working together manage to rise above what individuals can do in an almost mystical way. Small teams make transparency much easier, he says – they have a kind of collective awareness and can give things beyond what they are paid for. They learn brilliantly to deal with complexity and the unexpected. What might be possible on a larger scale? He and his colleagues were invited to explain their ideas to the American military at the Pentagon, but the conference was cancelled at the last minute.

Perhaps this is hardly surprising. Back in 1970, one study of the Pentagon by leading industrialists concluded that it was so hidebound by hierarchy that 'the astonishing thing is that anything works there at all'. This is the military version of what you might call 'Fantasy Efficiency Syndrome', and it can run very deep.

The idea of collective intelligence has begun to excite the cutting edge of organizational research. Consultant Robert Kenny described it like this: 'It's like something has shifted. People stop fighting for airspace and there's a kind of group intuition that develops. It is almost like the group as a whole becomes a tuning fork for the inflow of wisdom.' They look at how you mould egos into teams successfully, as the most successful American basketball coach ever describes. Phil Jackson turned the Chicago Bulls and then the Los Angeles Lakers into winning teams, he says, by being able to 'call on the players' need to connect with something larger than themselves'. Players are then able, somehow, to play beyond their individual capabilities.

'What's new today in the world is that now the first and most accessible gateway into deeper spiritual experience is not individual meditation but group work,' says Otto Scharmer of the Massachusetts Institute of Technology. But you don't have to buy into this as a spiritual phenomenon. Most people can point to experiences in their own lives when their team starts working together in a thrilling way. The point is that hierarchy, and controlling the thoughts and actions of individuals from the centre, is not the best way to make this kind of alchemy gel.

Radio was big in the 1920s and 1930s, and one of the places the new wireless sets were being constructed was in a Western Electric factory in the outskirts of Chicago, along with relays, capacitors and telephone equipment. At the time, every phone in the USA was made at Hawthorne. It was a big enough factory to have its own railway and hospital, and it also became the site of one of the most famous industrial experiments in history. It was also one of the first challenges to Frederick Winslow Taylor's time and motion ideas, just when they were at the height of their prestige.

The study lasted eight years from 1924, and was originally designed to find how much light affected productivity in the factory. A whole series of tests took place in different buildings to find what effect brightness was having. According to Taylor, there should have been one best way of doing it, but the results were frustrating and confusing. So confusing, in fact, that they brought in more researchers from Harvard Business School to repeat the experiment and find out what on earth was going on. The problem was that it didn't matter how strong the lighting was, even when it was little stronger than moonlight, but – every time there was a change – the productivity still went up.

When the Harvard tests began, they took six women who were studied intensively, testing every aspect of their work – even how much they slept, who they went out with and what they talked about. They asked them for suggestions and put them into effect. How about two five-minute breaks a day? Productivity went up. How about food breaks? Productivity went up. Shorter working days? Productivity went up again. Then, at the end of the experiment, they took away all the reforms, but – would you believe it – productivity went up again.

The true meaning of the so-called Hawthorne Effect has been disputed ever since, but most academics agree that the real reason the productivity went up was that the girls, and the workforce as a whole, responded to the attention. They felt more involved when they were asked for their suggestions. They felt less like cogs in huge hierarchical machines. In fact, none of the environmental conditions they were working in were relevant at all, despite what Taylor said. It was how they *felt* that was important. It didn't matter what systems or processes they had. Having their opinions asked made their work much more human. That is what made the difference.

Hawthorne is still controversial. Studies have pointed out that it took place partly during the Great Depression and the workforce knew their productivity would have to go up to keep their jobs. Even so, the basic revelation of the experiments was that human beings are not machines and that they respond to human connection. But it was also important that the women who were studied so intensively were working as a team. 'The six individuals became a team,' wrote one academic, 'and the team gave itself wholeheartedly and spontaneously to cooperation in the experiment.' Teams work.

The Hawthorne Effect is testament to the importance of human-scale management. It confirms that small organizations work better, primarily because people can bring their human skills to bear, but, even if the organizations are not small themselves, as the boundaries between companies begin to blur, the size of the units and teams people work in is enormously important too.

That isn't a new idea. Academics have been studying teams for well over a decade. The problem is that, despite the example of Gore and Semler, most organizations haven't acted on it. Those that have acted on it, have left the control mechanisms and hierarchies largely intact. That is why Rule 5 is to obliterate – not the companies or organizations – but the structures and systems that obstruct people working in a more human way.

Find out more

Seidman's *How* (see Rule 1) covers GE's Durham plant, but the key article was in *Fast Company* magazine ('Engines of democracy', Charles Fishman, 19 December 2007). A description of Professor Amabile's research on people's moods at work is in 'The power of ordinary practices' (Michael Roberts in *Working Knowledge*, Harvard Business School, 20 September 2006). Rosabeth Moss Kanter's research on self-policing is in her book *SuperCorp: How Vanguard Companies Create Innovation, Profits, Growth and Social Good* (Crown Business, New York, 2009).

See the references on Jack Welch under Rule 2. The phrase Management By Walking Around began with Tom Peters, as many of these ideas did, in his book with Robert H. Waterman, *In Search of Excellence* (Harper & Row, New York, 1982). The story about Asea Brown Boveri is from *The Individualised Corporation: A Fundamentally New Approach to Management* (Sumantra Ghoshal and Christopher A. Bartlett, Heinemann, London, 1998). There is more about the WL Gore approach to management in Gary Hamel and Bill Breen, *The Future of Management* (Harvard Business School Press, Boston, 2007). See also Rodd Wagner and James Hartnett, *12: The Elements of Great Managing* (Gallup, New York, 2010). As for Linden Lab, you can find the story – and similar ones – in Jeffrey Hollender and Bill Breen, *The Responsibility Revolution* (Jossey-Bass, New York, 2010).

The classic book about Semco is Ricardo Semler's own *Maverick* (Grand Central, New York, 1995). As for the Hawthorne Experiment, the classic description is in Elton Mayo, *Hawthorne and the Western Electric Company* (Routledge, New York, 1949). The main proponents of collective intelligence are mainly French, for some reason. The best website is undoubtedly Jean-François Noubel's at www.thetransitioner.org. His essay 'Collective intelligence: The invisible revolution' (translated by Frank Baylin, 2007) is available there. But there are also interesting links between this and how organizations work in *The Distributed Mind* (Kimball and Mareen Duncan Fisher, Amacom, New York, 1998). You can find some of the examples I give here in the special edition of *What is Enlightenment?* magazine, May–June 2004.

Rule 6

Give people whole jobs to do

When love and skill work together, expect a masterpiece.

(John Ruskin)

Only connect.

(E. M. Forster)

Summary

- Human beings can actually deal with complexity, but it is a different kind of complexity to the one that suits computers, the kind of instantaneous, sympathetic complexity of one human being dealing with another.
- The problem with inflexible systems is that many of the cases don't fit neatly into the approved categories – of course they don't, because people are different – and those cases just hang around unsolved, clogging things up.
- People need to feel a sense of achievement about the jobs they do, and that means they need to be able to see the whole job.

One long night in 2006, a repairman from the American cable giant Comcast arrived at the home of a man called Brian Finkelstein and, after some time on the phone, he fell asleep on the sofa. Finkelstein filmed him snoring and stuck it online, together with the sound track of a song called *I Need Some Sleep*.

The repairman was fired. But it transpired that he had actually fallen asleep after waiting over an hour on the phone to get through the useless systems that ran the call centre at his own office. This story seems horribly familiar to most of us who have to deal with organizations, and with call centres in particular. But there is something else familiar about it – the slow realization that it isn't the fault of the repairman, or the person on the end of the phone; it is the system, stupid.

In fact the poor individual in the call centre is probably as much a victim, if not more of a victim, than the people phoning up. They see only a tiny slice of the task that has to be done. They have to use a software system that often bears little relation to whatever the caller wants. They are expected to get rid of the caller as quickly as possible, are regulated about the precise time they are allowed to spend in the loo, and have every aspect of their work measured and reported to their bosses.

But it is the way their jobs, and so many others, have been salami-sliced that is important here. Call centres play a critical role in the design of the new customer interfaces, with their batteries of phones linked electronically to huge silos of back offices, which take the data from the call centres and make the work happen. The call centres face the customers, with their CRM software and scripts that appear on the screens in front of them. Then they chop up their requests and send them to different departments for processing. Somebody else will reassemble it all later. One of the most famous examples is the way our tax returns are now dealt with at HM Revenue & Customs. The number of people who deal with each return has increased from two to six – and every one of those handovers between them are opportunities for confusion, misunderstandings and mistakes (perhaps that is why a million people now pay the wrong amount of tax every year). The more work gets sorted, batched, handed over and queued, the more it has to be done again. We know that most medical mistakes in hospitals happen when staff hand over to the next team at the end of their shift. It is the same in offices, where nobody sees the whole job, except – theoretically at least – the distant manager, poring over the misleading statistics on his screen. They will be misleading, as we saw from Rule 2, because any statistics that are used to control people will always be inaccurate.

These clerical jobs have been chopped and diced as if they were a factory assembly line. Taylor and Ford's influence has spread over the century and, within 30 years of Taylor's death, the idea of the 'one best way' had become synonymous with efficiency in clerical work as well as factory work, as if manufacturing was the ideal model for everything from banking to doctoring. There is also a clue here about why human relationships are being taken out of the equation at work. They don't fit the assembly line model, where they are reduced to widget fitters in huge IT systems that take customer data and process them in little pieces. The problem is that splitting jobs up into tiny segments does not suit human skills, because the human ability to deal with human complexity – though not necessarily technical complexity – gets obscured. The result is miserable workers and rising mistakes.

It is worth a closer look at the distinction between these different kinds of complexity. Because the truth is that human beings *can* actually deal with complexity. Everything suggests that they do so all the time, and rather better than computers. But they are at their best when they face a different kind of complexity, the kind of instantaneous, sympathetic complexity of one human being dealing with another, factoring in a moment all that we know about people's past, their conversation, expressions, body language and feelings. People are not nearly as good at dealing with mechanical complexity, whether it is born of bureaucracy or IT or some combination of the two, which boils down the way people are supposed to act into virtual systems.

When IBM's mega-computer Deep Blue first won a game of chess against a human being (Garry Kasparov) in 1997, it forced us to face this distinction. No human being can factor in millions of different permutations and probabilities as well as a computer can. But they can deal with other human beings directly and simply in a way that computers and virtual and bureaucratic systems can't. Computers can write poems, yes, but they can't inspire or cajole like super-catalysts. Nor, incidentally, can they tell you much about what they meant by the poem they wrote.

So here we are acting out a kind of repeating tragedy as, every generation or so, those who run the world insist that every task is actually much more efficient if it was done factory-style. They do so despite the fact that, for the past half century or so, we have known perfectly well what a mistake this is. Back in the 1940s, the great American theorist of 'total quality' W. Edwards Deming warned that assembly lines, in themselves, are not efficient at all.

Deming's story is rather peculiar, because he found that his fellow Americans were not quite ready for this message, so he took his ideas to Japan after the Second World War and was enormously influential. Efficiency is all about getting things right first time, he said, because then you don't have to do it again. He was astonished at how much the American factory system wasted, in materials and time, just by failing to pay attention to quality. The result was the enormous sums of money were spent by organizations just to put right the mistakes they had made – and splitting up jobs means more mistakes.

'Let's make toast the American way,' Deming used to say. 'I'll burn, you scrape.'

Deming's idea of tackling the problem was to use 'quality circles' of staff. In companies such as Toyota, every member of staff famously had the power to turn off the assembly line whenever they saw something wrong. Three decades later, it was clear that Japanese industry was a good deal more efficient and effective than their American competitors. Yet it wasn't until 1978 that Deming began to have any impact on his own country. When executives from the American paper corporation Nashua visited the Japanese firm Ricoh, they complained later to their boss about the way their hosts had behaved. Ricoh had been so busy preparing for something called the Deming Prize that their visitors felt rather ignored.

'Prize?' said Nashua president William Conway. 'What the hell is the Deming Prize?'

It took until the following year for Conway to track Deming down and ask him to work for him. His first seminar at George Washington University attracted only 15 people. Ten years later, there were 50 Deming societies across the USA. Twenty years later, they were gone again. Quality circles were not an innovation that the IT consultants could earn much from, and quality management was quickly swept away by 'Re-engineering', and all the rest – as we will see in Rule 7. Those who believe that industrial assembly lines can be applied to all human work were once more in control, and with disastrous results.

Just as Conway was discovering Deming, the British industrial psychologist John Seddon was discovering the work of one of Deming's successors – Taiichi Ohno, the guru of Toyota's lean manufacturing system. Three decades on, Seddon is one of the leading systems thinkers in the UK, and he has done more than anyone else to take on this wrongheaded industrial thinking, applied where it doesn't belong.

By the 1990s, he was beginning the first of a series of bitter battles with those who tried to reduce human processes to neat industrial segments. He had been studying Deming at the same time as he was reviewing the first British quality standard, known as BS 5750, and couldn't see how the two related to each other at all. This revelation led him to take on the bureaucrats behind the international quality standard ISO 9000 (successor to BS 5750), convinced that it was making organizations less effective and less efficient by creating exhausting bureaucratic systems rather than getting people face-to-face. The standard actually 'encourages managers to act on their organizations in ways which undermine performance', he said.

Time after time, he found himself helping organizations that had signed up to IS0 9000, with its reams of paperwork and fearsome semi-trained inspectors, and found their service getting worse as a result. Far from increasing trust in organizations, or engaging staff in constantly making service better, ISO 9000 meant blind obedience to procedures. Companies such as the haulage contractor Eddie Stobart, which steadfastly refused to sign up – despite government instructions to do so – found themselves racing ahead.

In one company, Seddon found people made the sign of the cross when the quality managers arrived. 'The ISO campaign taught me that it was OK to question orthodoxy,' he told me. Ten years on, he is one of the most trenchant critics of turning companies or services into modern versions of Taylor's shop floor or Ford's assembly line, where employees are instructed to master one tiny bit of a job, in precisely the specified way.

This is an entrenched battle because, for well over a decade, the management consultants who advise the UK government all believed that public services should process jobs with this split between front office call centres and back office processors. From claiming benefits to buying computers, this is the system we usually find ourselves inside, from the moment we hear the first recorded message about which button we should press.

Seddon proposed an alternative, which sticks these two sides back together again, so that the experts can speak directly to the customers, so that each task is completed right first time by someone who can see it through from start to finish. He put this into effect in a series of companies, but then encountered the same problem in the public sector. More than a decade after his ISO campaign, his work brought him into a series of furious arguments with government auditors, first in New Zealand and then in the UK. The Audit Commission had become the defenders of the status quo, checking that local authorities and other government bodies were organizing their front and back offices in the approved way, and they became a symbol of everything Seddon had fought against. His furious monthly emails tracked his various engagements with them. But what really drew him into the public sector battles was his first experience of what happens when you sub-divide the task of processing housing benefit claims. The revelation was a result of being asked to advise Swale District Council in Kent in 2004.

Swale is based in Sittingbourne and they definitely had problems. There were 8000 people waiting for their housing benefit claims to be processed, the government fraud inspectors were due in for an unprecedented third visit, and the local MP was furious. Yet Swale had done everything they were told to do. They had split their system into two – separating a front office call centre from the experts in the back office

– but they were still bouncing along the bottom of the league tables. What was going wrong?

Swale was operating a system for processing housing benefit claims that was the result of a £200 million investment by the Department for Work and Pensions (DWP). People would apply by phoning the call centre. The call centre staff would send the applications for processing to the experts in the back office, and their every move was then measured – how quickly the phones were answered, how quickly letters were answered, how long the claim took to calculate, and so on.

In the case of Swale, Seddon found the real average was 52 days for people claiming housing benefits to get the money, but sometimes as much as 152 days, nearly six months, which is a great deal if you are a tenant – or a landlord, come to that. Seddon has worked a range of public services since, and he clearly relishes the moment he shows the real figures to managers. They often seem shocked, because they watch the target figures and they usually show something very different. Some councils which got four stars for excellence turned out to be no better than those that were doing badly. But of course this is so, he says – they were getting their four stars for doing it the wrong way.

In housing benefit, this is because targets measured the time between the claim arriving at the front office and the moment a decision was made. This was one of the government's 'best value performance indicators', but in practice it didn't measure the time it actually took for the average person claiming. Remember Goodhart's Law, which says that a target used to control people will always be inaccurate? That is because staff subjected to targets will always find ruses to make the figures look better. They kept the forms out of the statistics until they had been nearly finished, and then the inspectors would find out, add in more rules, and they would use their ingenuity to massage the figures some other way.

Seddon found they used one rule called 'nil-qualify', which disqualified claimants who had failed to produce all the information needed within a month. The case then closed and applicants had to start again from scratch, which stops cases which might break the target going through into the figures. In one council, he found that 40 per cent of claims were being nil-qualified. It meant that managers could say that their average length of time dealing with claims was 28 days, when it was actually 98 days.

Often this wasn't anything to do with the claimants, who were asked to bring in things again which they had already brought. The system was being used to protect frontline staff from the inspectors and to make their managers look effective, rather than what it was supposed to be for. It fed the delusions of those higher up the hierarchy that things were working when they weren't. In Swale, only 3 per cent of claimants were getting their claim settled in one call or one visit.

'We've forgotten our purpose,' one member of staff said. 'We're pushing paper to satisfy official specifications, not the claimants.'

Of course, there were targets about this to make sure people are seen within 15 minutes. In practice, managers often met this target by giving people a form and sending them home to get more information. When the queues began to gather outside from early in the morning, some managers in other local authorities hit on the idea of giving people numbers and letting the first 50 ask a question (only one, mind) while everyone else could only hand in documents. Asking claimants in the housing benefits queues, Seddon found people who were coming back for the fifth or sixth time, even sometimes the tenth time, bringing in yet more documents. The system was clogged up with its own processes and duplications, bouncing bits of applications backwards and forwards between front and back offices.

The system sounds insane, but it is used in one form or another in many businesses and public services. It means that the front office tends to fragment jobs, which then have to be put back together again by the back office – without any human contact with the person who is claiming, and feeling no responsibility to them. If the case is 'incomplete', they have no qualms about taking it out of the target statistics, sending it back to the front office for more information, and so the volume of work grows – a huge machine for creating work, at vast expense. The targets are met but the service is appalling.

To stop fraud in housing benefit, either by the claimants or by the front office, the DWP began asking back office staff to use, what they called, a Verification Framework. This was a checklist of information they would need before any claim could be processed. Yet in practice it meant more room for misunderstanding, because the back office experts took a different view about how much of the Framework applied to the human being on the end of the phone. One side was using their human skills, despite all efforts to root these out, and the other was applying a set of rules. The Framework was designed to be inflexible, so of course it failed to deal easily with lots of different kinds of people. Or as Seddon says, it stopped the system from 'absorbing variety'.

After eight years, the Verification Framework was abolished, after the DWP had spent nearly £320 million on it, largely on IT. It had become clear that it was actually raising costs.

It was actually the inflexible system that was making things so ineffi-cient, and this was Seddon's great discovery: the key problem about systems that are inflexible is that many of the cases don't fit neatly into the approved categories – of course they don't, because people are different – and those cases just hang around unsolved, clogging things up. Staff who are trained not to use their initiative just pass them on elsewhere, and the poor people keep phoning up and the work mounts up. Seddon called this 'failure demand'. He meant all those people contacting the call centre because something has gone wrong, or because they haven't had a reply – all that extra work which an effective system shouldn't have to do at all.

In Swale, he discovered that up to 78 per cent of calls were not from people calling with new claims but from people finding out what was causing the delay. This added to the workload and provided him with the big idea: that there are two kinds of demand on call centres or public services – real demand and failure demand. Real demand has to be catered for; failure demand needs to be diverted. But most organizations don't distinguish between the two, partly because the experts are divided from the people who are calling. Or, worse, because they are paid according to the number of calls they get.

Seddon's answer, which he developed working in the private sector, is to bring the front office and back office together again so that people can speak directly to the experts with the power to do something about it. That meant the system could be flexible enough to solve people's problems there and then. It would give staff the responsibility to do the job in the way they know best, and to judge that according to their own criteria.

Seddon managed to reduce Swale's time to process housing benefit claims right down to five or six days. Within five months, they had gone from one of the worst in the country to one of the best. They now boast that they are 'dealing with the claim as it comes in', which is the way to do it. The trouble was that other local authorities that made a similar shift – ending the split between front and back offices – then risked being rapped over the knuckles by the Audit Commission. The inspectors demanded that councils show them plans for 'shared services' with other local authorities, and expect to check out aspects of a system which were no longer there.

The problem was that the government relied on experts in the management and IT consultancies who were completely committed to the same split. They were not exactly open to Seddon's message.

'I came to the conclusion eventually that I was wasting my life,' said Seddon. 'The next time I got a call from the government, asking if I would come and see the minister, I said he can come to me or talk on the phone. But I still spend a lot of time helping managers argue with the Audit Commission. It has become a passionate hobby because it matters so much.'

This particular rivalry exploded during the summer of 2009, when the Audit Commission finally roused itself to bite back. Seddon had been goaded into a fit of radicalism when he heard the Commission chief executive Steve Bundred say that lower pay for public sector workers would be a 'pain-free way to cut spending'. 'Politicians often wonder why it is that their local services receive four-star ratings, yet their surgeries are full of people complaining,' Seddon replied in *Local Government Chronicle*.

The Audit Commission's communications director David Walker retorted that Seddon was being 'nonsensical'. 'To him all organizations seem to be the same: working for Toyota is equivalent to working for Torbay,' he said. 'He seems to live in a world in which there are no taxpayers, anxious about value for money, pressuring ministers who in turn seek assurances that public money is being effectively spent.'

Actually, the Commission got it wrong (it was scheduled for abolition the following year). Seddon's message was almost the opposite of this – that offices or surgeries and schools are not the same as car factories. Toyota, or 'Lean', just doesn't work where you need human skills.

Why Seddon's crusade is so revolutionary, and so suspect in some quarters, is that he implies there is a vast amount of waste in the system, much of it caused by the targets concocted by what he calls the 'Regime'. He holds out the opportunity of enormous savings in the public sector, just by putting front and back offices back together again, giving responsibility to front line staff and connecting them in a human way with the people they are supposed to be helping. Failure demand isn't small. Seddon believes the proportion of demand on call centres in financial services which is just the result of failure elsewhere can be anything from 20 to 40 per cent. In utilities and local authorities, it is sometimes as much as 80 per cent, or just occasionally more. That in itself has frightening implications for the cost of public services.

Take one of Seddon's recent projects, to improve the antisocial behaviour unit at Hounslow Homes. The original system was designed on the approved basis, dividing front office call centre from back office experts, and it was meeting its target of 100 per cent of calls responded to within 24 hours. In practice, Seddon's team found that three quarters of these calls were actually from people chasing officers to find out why nothing was happening. They discovered that officers were meeting the target by calling people back, even if they couldn't get through, and leaving a message – so people had to call again. It could take two or three weeks just to speak to someone who wasn't just managing customer management software. Cases took up to two years to actually solve.

'We were always ticking boxes and the focus was on meeting targets rather than doing what matters, and we lost the customer in the whole thing,' explained estate manager Sekandar Ravi. This is a familiar problem, in the public and private sectors alike. But with the help of Seddon's team, they designed a new approach without targets, procedures or standards, and which takes calls directly from the public, and where officers could listen to them without following the prompts on the screen from their software. And above all, where they can get out of the office and meet people face-to-face.

It also seems to work. 'Now we are getting real engagement, eye-to-eye contact and a firm handshake,' said Ravi, 'it feels that we have actually done something and it feels good.' The volume of demand has plummeted and dropped sharply. There has even been one instance of a satisfied customer bringing chocolates in for the staff. The system works, says Jill Gale, the director of housing. 'We have re-humanized it.'

Seddon's inspiration came partly from Deming and partly from the man who created Toyota's 'Lean' management system, Taiichi Ohno. The Toyota way of doing things dates back to the immediate post-war years when Ohno and his team visited Ford to see his assembly line at work and were not very impressed. There was too much extra work from bad workmanship, and there was far too much extra stock lying around taking up space. On the same visit, they happened to shop in a supermarket called Piggly Wiggly and then they got excited. The supermarket had a system that only ordered enough stock as it was bought by customers. It was the beginnings of the Just-In-Time delivery systems that so many companies use today, and the basis of Toyota's success.

Toyota's 'lean' systems are all the rage in the UK now. Companies such as Tesco use it to organize their massive distribution system. Public service administrators, fearful of the spending cuts, are increasingly looking towards 'Lean', and it was famously used in Bolton to revitalize an inefficient pathology department, speeding up the business of doing lab tests just by changing the positions of the machines and a range of other small reforms. It tries to bring jobs back together again rather than splitting them all up into little bits. With its emphasis on teams and quality, and on Management By Walking Around and seeing for yourself, 'Lean' has a great deal in common with some of the innovations in this book – but it is a distant relative, rather than a close one.

The problem with 'Lean' outside a factory is that it keeps the Toyota assembly line at its heart. It requires standardized systems and processes. It means measuring the speed that everyone does their jobs and holding them to the new system. It is therefore a prime candidate for putting the processes into inflexible concrete, and 'Lean' IT is the latest twist in what is on offer from IT consultants. The trouble is that 'Lean' assumes everywhere is a factory, a little like Toyota. The central purpose may be to improve the flow by bundling jobs back together again, but it isn't necessarily about getting people back face-to face – and it certainly doesn't mean letting staff get on with their jobs in the way they know best. Quite the reverse.

'Lean' techniques will help make an organization more efficient, by improving all the processes, but it won't necessarily tell you that you need a completely different organization. It will improve the flow through the pathology department, but it won't tell you if you really need a pathology department. It won't tell you if you need a different kind of NHS altogether. That's the problem with 'Lean'. It still doesn't put human beings right at the heart: it will make processes more efficient, but not necessarily more effective.

But what we do need to learn from Ohno and the Toyota way is the crucial importance of knitting jobs back together again. Splitting jobs up means that people feel less responsible for getting them right. They are that much less likely, from experience, to sense that human being behind the job who needs them to take trouble over it. Splitting jobs up into separate processes, or experts from the people they are supposed to be helping, is assembly line thinking after the end of the age of the assembly line.

Real efficiency means minimizing the amount that jobs are chopped up to different people and getting the human beings to connect. People need to feel a sense of achievement about the jobs they do. That is only human. But to feel anything like that means that they need to be able to see the whole jobs. Face-to-face jobs may cost more money, because there is less automation, but it certainly cuts out waste. In the end, effectiveness is always more efficient than ineffectiveness, for the reasons Deming always said – you don't have to scrape the burnt bits off to make toast.

The strange thing is that this is not a new insight. While Ohno was relaxing into a well-earned retirement (he died in 1990), a similar idea was emerging in the mind of one of the most senior commanders of the United States Air Force (USAF). In fact, it was this insight that would soon make General Wilbur Creech, in command of Tactical Air Command, into a byword for a new style of organization. It was 1978 and the USAF was suffering from a technical crisis. They had 3800 aircraft, but only half of them could fly at any one time because of chronic mechanical problems.

Creech was a former Vietnam War fighter pilot, and had been in charge of electronic systems for the USAF before taking up the post. He decided that part of the inflexibility came from breaking jobs down into tiny bits. One part for a F15 fighter meant 243 entries involving 22 people and 16 hours on the administration and record-keeping alone, even before a new part was made or fitted. He also realized that there was a damaging disconnect between the pilots and the technicians and mechanics who were serving them from a central unit. The USAF was, in short, organizing itself in much the same way that management consultants have been urging the UK government to organize itself a generation later – shared central services, fragmented jobs, central purchasing.

Creech's solution was to assign the mechanics to the squadrons, so that they could have a direct relationship with the people they were serving. He decentralized the supply of parts and planes, and gave each squadron control over how much flying they did, and dumped all that nonsense about economies of scale. When he left in 1984, 85 per cent of planes in the USAF were operational at any one time and the improvement had been achieved without any extra money or people.

'We think it was organization,' said Creech. 'We think it was decentralization. We think it was getting authority down to the lowest level. We think it was acceptance of responsibility to go with that authority. We think it was a new spirit of leadership at many levels – making good things happen.'

He went on to write a book called *The Five Pillars of Total Quality Management*, a homage to Deming's ideas, and became a business advocate of putting people at the heart of organizations. The human dimension is far more important than process and systems, said Creech. 'If 90 per cent of employees believe that productivity is in their best interest, productivity will rise. It's time for us to pull out the old habits that were good at another time, but are no longer working.'

Creech was one of the key figures behind Al Gore's innovative National Performance Review in the 1990s, which managed to embed many of these ideas into American frontline services. They did not cross the Atlantic.

John Seddon's furious newsletter recently included a letter from a local government officer complaining about what happens when the system drives customers and experts apart. This was the sad story of their IT helpdesk. Once upon a time, they used to phone up the IT department on the next floor and ask for help. The IT experts would either fix the problem over the phone or agree a time to pop down. But all that changed when they got an approved central IT department for the whole council instead. Now, when staff phone up for help, they are not allowed to speak to an IT specialist directly. The helpdesk assigns them a job number and says they will be in touch (they can't say when). 'Unfortunately, I have meetings to attend and guess when he comes,' he wrote. 'So it's back to the helpdesk.'

> 'Hi, one of your staff came when I was out today.'
> 'Have you got a number?'
> 'Yes, it's 23563.'
> 'That's right. He is scheduled to come out to you today.'
> 'Yes, I know. I had to go to a meeting and missed him.'
> 'Do you want to make a new report then?'
> 'No. Can I arrange a time for him to come?'
> 'No, I can't make appointments. But I could make a new report and give you a new number.'

This is the system approved for most of the public sector in the UK and great swathes of business. It is deeply frustrating and ineffective and it is a testament to what happens when you sacrifice face-to-face relationships on the altar of economies of scale. The answer is to stick systems back together again, give people whole jobs to do because – in the end – that is what human beings do best.

Find out more

The key text here is John Seddon's *System Thinking in the Public Sector* (Triarchy Press, London, 2008). His websites keep the argument fresh, especially www.thesystemsthinkingreview.co.uk and the website of his consultancy Vanguard, www.systemsthinking.co.uk. Another interesting source is the government's own report into systems thinking in housing (also largely written by Vanguard) which is called *A Systematic Approach to Service Improvement* (ODPM Publications, Wetherby, 2005).

Frederick Winslow Taylor is also very relevant to this chapter, and his best biography is Robert Kanigel's *The One Best Way: Frederick Winslow Taylor and the Enigma of Efficiency* (MIT Press, Boston, 2005). Some of the most useful coverage of Deming and total quality I found in publications from the 1990s – for reasons I discuss, Deming has gone out of fashion among the mainstream gurus of efficiency. See for example

Robert R. Locke's *The Collapse of the American Management Mystique* (Oxford University Press, Oxford, 1996) and John Seddon's *In Pursuit of Quality: The Case Against IS0 9000* (Oak Tree Press, Dublin, 1997). There is more about Deming at the website of the W. Edwards Deming Institute (www.deming.org) in the USA and the Deming Forum in the UK (www.deming.org.uk).

The story of 'Lean' thinking in the pathology department is from Daniel Jones and Alan Mitchell, *Lean Thinking for the NHS* (NHS Confederation, London, 2006). The example of General Creech comes from a key text in this area *Reinventing Government: How the Entrepreneurial Spirit is Transforming the Public Sector* (David Osborne and Ted Gaebler, Penguin, New York, 1992).

Chuck out the big IT systems

I'm not a big emailer, though; it's a crutch that hinders person-to-person communication.

(Howard Schultz, CEO of Starbucks)

Computer says no.

(*Little Britain,* BBC)

Summary

- Corporate software has ushered in an era when customers are very much less important than processes.
- IT could be used to give power, responsibility and independence to employees, users and customers alike. Instead, it has been used to measure the unmeasurable, and to use those measurements to control.
- We need to build face-to-face relationships and develop software that genuinely sets us free to do so.

When the journalist Simon Head visited the Nissan engine and assembly plant in Sunderland in January 1994, he had a revelation that explains how too much measuring power, and too little people power, had led to the very opposite trends at work to those we had been led to expect in the 21st century.

Head has specialized in writing about changes in the way we work. He had been fully exposed to the full rhetoric of American IT theorists, who argued that expert systems could enhance people's skills and give people responsibility again. They could use data to get a complete picture themselves and act on it. They could solve problems and work in devolved teams. IT meant a new era of skilled, empowered employees and an end to the repetitive assembly lines that the journalist Horace Arnold described at Ford a century ago.

We all know that it didn't happen quite like that. Having your boss on the end of a mobile phone or email meant that you were often less free than before. Managers often prefer to take all the decisions if the technology lets them. Because managers *can* take more decisions, then in practice they do – or they wait for their own superiors to do so. But the full truth about what was actually happening only became clear to Simon Head when he went into the Nissan car plant that had become a symbol of modern industrial Britain:

> For there before me was a cleaner and no doubt less noisy version that Horace Arnold had described 80 years before at Ford's Highland Park plant. Hordes of young workers swarmed over scores of cars as they moved slowly along the assembly line. Workers performed the same simple tasks over and over again, and there was a palpable sense of stress as workers struggled to get their tasks done within the amount of time it took for the vehicle to pass through their segment of the line.

Every worker had with them a little blue book that set out in intricate detail exactly what movements they had to follow for each process and recorded how they had managed to do it in time. These were not skilled workers with new powers and responsibilities delivered by IT. In fact, like other Japanese car makers, Nissan had a policy of shrinking the number of skilled trades they employed. What was actually happening was that the controlled time-and-motion world of Frederick Winslow Taylor was being bundled up into new software, which was creating a new straitjacket for employees, not just in factories but in offices too.

It was a peculiar and contradictory time. Former McKinsey consultants Tom Peters and Robert Waterman published their groundbreaking management book *In Search of Excellence* in 1982 and it had set out a revolutionary people-centred strategy for business, the very opposite of what their McKinsey colleagues were pedalling. It was full of words such as 'listening', 'empowering' and 'ownership'. It permanently shifted the kind of language that businesses use, and dovetailed nicely with the parallel search for 'quality'. From then on, no corporate executive could avoid phrases such as 'our people are the heart of what we do'. Peters has continued his maverick career railing at the crass stupidity of corporate strategy and systems, especially where it leaves out the human element.

The new emphasis on people and leadership has raised top corporate salaries – if people are not interchangeable, they need to be paid better. The trouble is that it has usually remained rhetoric, and for some reason has still not filtered into the huge organizations in the public sector. The other problem was that it provided the consultancies with nothing to sell (nor did Deming's quality circles). Bundling up Taylorism and software was something different: that could be sold as a profitable product.

This combination was potentially lucrative for the consultants, but it required a big idea behind it before it could be sold. That came along in 1993 with the publication of the bestseller *Re-engineering the Corporation*, by mathematician Michael Hammer and computer scientist James Champy. Hammer had begun what became the 're-engineering' revolution with a typically aggressive article in the *Harvard Business Review* three years before called 'Don't automate, obliterate'. It was a management philosophy for the software age and it made a great deal of sense. It denied that there were such things as economies of scale. It said that it made no sense to split up the functions of a company and do them separately. They needed to be brought together in one system, and that meant that middle managers had no real function left.

Re-engineering meant huge redundancies. A third of BT staff lost their jobs in the mid-1990s, which had a massive impact on morale. It also carried within it a major IT problem. Re-engineering was supposed to make organizations more flexible, but it didn't happen that way. When the software was written that was capable of drawing all these various organizational functions together, most companies were tempted to split their operations into two – as we saw in Rule 6, there would be a front office factory of untrained call centre staff facing outwards, and a back office of experts who would take decisions and make things happen. It was as rigid a structure as any of the departments they had all replaced.

While Hammer and Champy were putting the finishing touches to their book, the first enterprise resource planning (ERP) software was put on sale by the German software giant SAP. Their English version of ERP R/3 was available in 1993, and by 1994 it was earning more than $1 billion. ERP dovetailed with re-engineering. It offered to pull together all the functions of an organization in one management system and the big corporates loved it. But by the end of the decade, SAP and rivals such as PeopleSoft and JD Edwards were finding that the boom was already over. Some of them were losing money on ERP and they were beginning to shift their efforts into selling Customer Relationship Management (CRM) software instead, to regulate precisely what call centre staff could say in any given situation.

ERP delivered by consultants was equally depressing. A 2000 report showed that 92 per cent of American companies were disappointed in their vast investments. It was the same in Europe. Companies carried on with the idea, but did so increasingly by buying software off the shelf from SAP or Oracle, which meant it was often even less suited to the particular needs of individual companies, still less to those of individual members of staff. The management consultancies had a moment of panic and then looked towards the public sector. That was how IT re-engineering reached into the huge budgets available to the world's governments.

The consultancies had been wooing governments for some years, and particularly in the UK. The future Labour minister Liam Byrne, an Andersen Consulting consultant, had organized a training session for 100 Labour MPs a year before Tony Blair swept to power. It was hardly surprising that the new ministers in the Blair government saw Andersen Consulting (now Accenture) as a source of special advisors, or for independent advice to bolster their arguments against their civil servants. Soon the senior civil servants were becoming interchangeable too. The chairman of the Inland Revenue went to help PricewaterhouseCoopers clients minimize tax. The Treasury's managing director went to KPMG.

It so happened that the British government's attention was elsewhere in 2002, nervously planning for a controversial war in Iraq. They were also frustrated that their public service reforms seemed to be bogged down – the targets were coming due – and the consultancies swept in to fill the vacuum. By 2004, the UK government was spending a massive £25 million a day on management consultancy. In the following year, total spending on consultants for the Ministry of Defence leapt from £83 million to £148 million, and the NHS from £25 million to £85 million.

It was no coincidence that this was also the first year of the most disastrous government investments in IT systems. Many of these were related to ERP, dividing heavily monitored frontline call centre staff from the experts in the back office who would process decisions. The huge tax credits IT system, bought from Ross Perot's old company Electronic Data Systems, managed to pay out £2.2 billion too much in its first year. During that same period, there were 100 million calls to the tax credits helpline, half of which went unanswered. That was the pattern.

'Chronic dependence on consultants is an implicit admission of ineptitude in management,' wrote a former senior executive at the telecoms company AT&T, which managed to fritter away half a billion dollars on management consultancy as it slipped slowly down hill in the early 1990s ($96 million to McKinsey alone). What it actually revealed, as it does with a similar dependence among UK ministers and civil servants, was a lack of confidence in their own abilities. It may also be a vicious circle. If the solutions that most management consultancies provide don't work as well as they should – a mixture of Taylorism, IT re-engineering and command-and-control centralization – it leads to even less confidence and more money frittered away, and so it goes on.

So it was that the targets culture, which had been ushered in on both sides of the Atlantic by McKinsey and others, was welded onto a related culture of corporate re-engineering. It was powered by enormous IT investment and massive call centres, regulated by ERP and CRM software. IT can be used to enable human relationships, but in practice that isn't what happened. It was used to divide people – professionals from customers, experts from people with problems – because it looked efficient.

This is how it still works. The software teams look in detail at all the processes, find the best employees and watch what they do. Then they turn that into the processes staff are led through on the screen. They also build in onerous reporting, tick-boxing and measurement systems to satisfy managers that the performance of every part of the process can be measured. Managers want to be able to stare at a complete picture of the machine in motion until they can find who to blame if it isn't working, and ERP gives them this illusion, as we saw in Rule 2. 'These are the assembly lines of the digital age,' wrote Simon Head, 'complete with their own digital proletariat.'

There was no organizational form, no new skills, no self-managing teams and no new kinds of decentralized working, at least not when it was delivered by re-engineering software. If IT investment was really about empowering ordinary employees, then you would expect them to value their middle and lower income staff as experts in the frontline, said Head. In fact, they were de-skilling them and, where possible, making them redundant.

People such as John Seely Brown, a former director of the IT think-tank Xerox PARC, saw what was happening as early as 2002 and described it as 'technologically inspired vandalism'. What was emerging instead was a series of 'monolithic blocks of concrete', where the accumulated experience of staff, and their ability to make human relationships – even brief ones over the phone – were being lost.

The complicated rules, enforced by the software that staff were using, were ushering in an era where customers were very much less important than processes. The rules create an intractable combination of processes, Kafka-esque customer service and a peculiar belief that what comes up on the screen is real.

Strangely, this whole development was predicted by one of the fathers of British systems thinking, Stafford Beer, who wrote an essay in the mid-1970s, which described organizations even then as 'grotesquely top heavy and ruinously expensive'. Beer saw clearly that IT would be misused to entrench these problems, taking an organizational structure that no longer works only to 'enshrine it in an expensive solid state sarcophagus'.

'It knows nothing,' he wrote. 'It will countenance nothing that it does not know. It can't move. It is death.'

Then there was the NHS computer project, the apotheosis of the idea that systems work better than people. It was to become subject of a series of inquiries and panic measures, as costs ballooned from £2 billion to well over £12 billion.

The project began with a meeting in 2001 between Microsoft founder Bill Gates and Tony Blair and his closest entourage. Gates' enthusiasm inspired Blair, who was always rather a pushover for a technological solution. The programme was launched in June 2002 and christened 'NHS Connecting for Health'. It was to be the biggest non-military computer project in the world. Those in charge could have opted for local decision-making guided by national standards. Certainly, the report they commissioned from McKinsey said that no existing contractor had the capacity to take it on. But Whitehall has a deep-seated fear of local decisions, so in the end Connecting for Health was to be planned as centrally as possible. So it was that, in the spring of the following year, the NHS put an obscure advertisement in the *Official Journal of the European Union*, alongside notices about huge EU contracts for missiles or motorways, and the project was born.

It wasn't as if there had been no IT in the NHS before. But Connecting for Health was like starting afresh, and adding a great deal more to control the way staff did their job. There was to be – and may still be – an online booking system, centralized medical records for 50 million patients, e-prescriptions and computer links between NHS organizations, all to be ready by 2012. A handful of huge IT companies hunkered down with the contracts, even when they had been turned down, because the smaller firms which won them couldn't cope with projects on that scale. NHS trusts found themselves instructed to upgrade by their bosses, but they weren't allowed to see the contracts to see if the prices were justified (they were 'commercially confidential'). At the height of the NHS funding crisis of 2005, some were having to transfer money away from frontline care to pay for the IT project.

The first signs of real panic came in 2007 when the computerized system to allocate new doctors to hospitals screwed up spectacularly. Among its glitches was assuming the whole of south-east England was one place, so that it happily allocated husbands and wives to hospitals up to 100 miles away from each other – when they allocated them at all. Something was going wrong. It was about this time that one of the consultants, Fujitsu executive Andrew Rollerson, poured cold water on the project. 'There is a belief that the national programme is somehow going to propel transformation in the NHS simply by delivering an IT system,' he said. 'Nothing could be further from the truth. A vacuum, a chasm, is opening up.'

Not surprisingly, Fujitsu abandoned the project altogether soon afterwards, complaining that they had been asked to change the specifications 650 times for no extra money. Accenture went the same way. By the end of 2008, not a single health trust had been supplied with the new Lorenzo records software. The 2009 crash at consultants CSC cut access from 80 hospitals to all their IT systems for four days, the biggest IT failure in NHS history. The Choose and Book appointment system – which stops GPs sending patients to specific consultants – kept crashing and diverted patients to hospitals 50 miles away or more. One in 16 calls were not answered at all, and most that did get though heard a standard message saying that no appointments were available.

Nor was Connecting for Health the only vast IT project that was causing concern. There was e-Borders for the immigration service, and the huge ID card project, ContactPoint, the controversial database of every child in the country, cancelled by the coalition government in 2010. The national offender management system, C-Nomis, proved to be impossible and was cancelled as the costs had tripled. The cost of the tax credit computer system shot up from £3.5 billion to £8.5 billion. The online passport application system was cancelled because the software would have been obsolete by the time it was finished. At one stage, the Office of Government Commerce began to panic and ordered staff to shred documents about government IT upgrades.

Why should Britain be quite so prone to IT cock-ups? A study for the London School of Economics accused some firms of pitching prices low to start with, knowing that the civil servants would change their minds constantly and the eventual contracts would balloon four or six times over. But, governments being what they are, that probably applies everywhere.

Even so, there does seem to have been a fatal naivety about the British approach to IT, partly because the politicians have so little idea what is possible. Partly also because British ministers, consultants and managers have been more in the grip of this narrow kind of efficiency than almost anybody else. Government is bigger than most businesses after all, and that much more difficult to control. It made ministers keener on big, controlling solutions, and less sceptical about the control it claimed to give them. Nor did they seem to understand the limitations of centralized databases, and how vulnerable they are, nor the way databases degrade. A third of entries in the vast Driver and Vehicle Licensing Agency database have at least one mistake. The Connecting for Health patient contact service already has hundreds of thousands of records with the wrong addresses and wrong GPs, so letters go to the wrong places and patients don't turn up for appointments.

IT consultants talk about the 'coffee machine phenomenon' at ministerial briefings. After listening politely to government decision-makers outlining their latest IT ambition, the possible contractors meet around the coffee machine and agree with each other that the scheme is impossible – then they go right back to listening politely again, all the way to the contract stage. 'In the public sector, you get this willingness to grind onwards towards the abyss,' wrote Richard Bacon, one of the MPs most concerned with making the NHS IT disaster public. Even when senior officials at the Rural Payments Agency said that the huge new IT system for farm payments would be a nightmare, they went right ahead with it anyway – and it was.

It was a strange world, and we still live in it. Because governments and consultants alike have misunderstood what role IT ought to be playing. It could be used to give power, responsibility and independence to employees, users and customers alike. Instead, it has been used to measure the unmeasurable, and to use those measurements to control.

In 2007, a student called Paula Ceely took her satellite navigation equipment so seriously that she drove onto the main line in front of a train (she heard a train coming, scrambled clear before the impact, and blamed her navigation equipment). That same year, schoolchildren from Fareham in Hampshire spent all day in the coach looking for Hampton Court Palace, frustrated by a driver who insisted that they follow the sat-nav instructions to Hampton Court Lane in Islington. It is this same blind faith in what the IT equipment tells you that has so affected the judgement of managers.

This is a small part of the widespread idea that data is somehow all you need. The UK National Probation Service increased its budget in real terms by a fifth up to 2007, but employed 4 per cent fewer probation officers. Most probation officers have a caseload of at least 60 individuals, as if somehow a hugely expensive computer programme would allow them to monitor the performance and behaviour of every former prisoner in their care. As if it was a shortage of data they suffered from, rather than a shortage of super-catalysts capable of making relationships with people who might want to change their lives.

It was the same with the American company Educational Testing Services, which assumed they just needed a good system when they disastrously took over SATs testing for UK primary schools in 2008. Hardly surprising, then, that markers didn't get their papers in time, papers were sent back to schools without having been marked, and they had to take on extra staff to deal with 10,000 unanswered emails as the deadline loomed.

It was the same illusion that all Intel has to do is to wire up the homes of older people to measure their movement patterns and predict medical problems. As if Matsushita's 'smart toilet', which automatically analyses our urine each day and sends it automatically to our doctors, is going to help us stay healthy. As if making 400 staff redundant by email, as the American retailers RadioShack did, was really more efficient and caring. The information is only part of the problem here. Who is going to act on it? Who is going to go round to the old people and help them stay healthy? Who is going to cope with the deluge of patients acting on an alarm from their toilet? In short, we have an information overload, but a serious shortage of ideas about who is going to act on it.

Worse, we have entered a world where there are American software systems that mark essays automatically – by counting words such as adjectives, connectors and indicators, adding up the 'howevers' and 'moreovers' to see if there is an argument – but ignore the content completely. One San Francisco company, Cataphora, helps employers evaluate staff on the basis of whether their emails are forwarded to others, because they must be ideas-generators.

There are many things that IT systems give us that we couldn't possibly do better ourselves. Indexing, databases, counting, stock-taking. I am writing this book using word processing software, and I certainly would not be without it. What IT can't do, and should not try to do, is to replace the tasks that human beings will always do better, because they require some kind of relationship.

Take, for example, my son's latest school report. His primary school is brilliant, and in many ways primary schools are the jewel in the crown of public services in the UK, simply because they remain committed to human interaction and human scale. But even here, some of the phrases are bizarrely bland. 'He is developing his understanding of the numbers to 20' (well, it is maths). 'He has required support to understand that labels carry key pieces of information' (what on earth does that mean?). The real problem, at least in the blandest reports, is that these phrases are rarely written by the teacher. They create their reports using software such as ReportAssist or Teachers Report Assistant where they tick the boxes related to attainments and pre-set phrases pop up. This also explains the banality of so many Ofsted reports about schools. The answer is that the human element has been removed. The inspectors tick the boxes for specific standards and the phrases go straight into the report.

That is the problem: we are not interacting with a human being, but with a computer programme, and are therefore not getting the insights we should, at least in such a way that we can act on them.

Of course report-writing software saves time, but it also makes the exercise all but pointless. It is a small example of the way that IT has been used – at vast public expense – to hollow out our institutions, rather than to support professionals to do their job more effectively. It has been used to standardize and control, not to empower. It has encouraged the idea that people are interchangeable, that human skills are endlessly replicable and that the systems matter more than the customers.

This is why the Japanese railway company Kelhin Electric (now Keikyu Corporation) introduced a computer scanner in 2009 to check their staff's expressions each morning, to make sure they have the required cheeriness. It is why customers say they have to spend an average of 23 minutes on the phone to BT to sort anything out.'Overworked staff are routinely ignoring mistakes, refusing to answer the phone calls and even binning letters,' said a whistleblower from HM Revenue & Customs. 'People who ring in with complaints are cut off simply to meet government targets on answering calls.'

These issues are disputed because they are about what it means to be human, and whether you believe artificial intelligence can ever quite replace human intelligence, whether there is anything human beings can do – such as having genuine human relationships – that can never be replicated by a machine. That is why it is rather sensitive. There are tidal waves of hidden argument underneath. But a relationship with a shiny virtual construct is a metaphor, no more than that. If we trust the occasional brand name, it is because we trust the people behind them. If we meet some of their representatives occasionally, that trust can go deeper. But we still have to interact with them. It certainly is possible to build a virtual relationship with a brand – as we do with Amazon when they recommend books to us – but don't let's pretend that any of this is real. Anyone who thinks they are having a real relationship with Amazon, rather than being plugged into a sophisticated database, needs to get out a bit more.

So the answer is to be brutal about this – chuck out any IT system that tries to reduce, define, standardize and control. We have to escape from the straitjacket set for us by Frederick Winslow Taylor and his successors and make sure we don't waste the skills of the people we employ. We need to build face-to-face relationships, and develop software that genuinely sets us free to do so. That is what works.

Innovative businesses have been able to grasp the basic problem, and realize what we lose when we take out the human element. 'People can see something dodgy that a machine might miss and if they phone someone, they can often detect something in their tone of voice, or whatever, that a machine cannot do yet,' said PayPal's head of risk management Garreth Griffith. 'Some things we want to do manually, even though it slows the process down, because there is definitely an element of intuition about it.'

The American transport company Penske is another one. 'A lot of jobs in the future will deal with face-to-face interaction,' said Stephen Pickett, Penske's chief information officer. 'You can't do a process analysis over the phone, you can't understand the inner workings of a corporation over the phone. You have to understand how a user wants to use software. These are face-to-face jobs, feeling the good times and bad times, knowing enough about the company.' That sounds a little backward-looking, but it is actually the future.

The business of ordering a round of drinks in an Edinburgh bar during happy hour stayed with John Timpson, and informed his attitude to the way frontline staff are usually controlled by IT systems. The barmaid sold him two drinks for the price of one, but she only brought one over to his table. 'Sorry,' she said. 'The computer shows 7pm – it won't let me serve the second drink.'

Timpson was always sceptical of IT and the information it feeds through to managers. This was reinforced when he finally bought his rivals Mister Minit in 2003 and found that the system distributed the wrong stock and fed the wrong information through to head office. He believed it contributed to the loss of £120 million over two years, not just because of the mistakes, but because fine-tuning the system took up so much of the marketing department's time. People still use computers in Timpson's – they even sell products on their website – but they are not linked together in such a way that IT gets between the frontline staff and the customers. If they want to sell someone two for the price of one, the EPOS system on the till won't stop them, because there isn't one.

When Timpson bought Mister Minit, out went their EPOS system. It was the same when they bought Automagic in 1995 and their other acquisitions since. He now says he is paranoid about EPOS: 'Their systems had given all the power to head office and turned shop staff into robots ... It isn't always right to copy everyone else – just ask the banks. We have grown as much as we have by giving our colleagues freedom to order their own stock and look after customers the way they know best.'

Like Timpson, Semco also threw out its mainframe computer system during the 1990s and wound up their IT department. There are lots of computers, of course, but they relish the freedom of using IT to help them take more control of their work, rather than the other way around.

Ever so slowly and haltingly, some of this seems to be filtering into public policy. It is happening despite the best efforts of the IT consultancies (the Gershon Committee in 2004/2005, which urged ever-faster adoption of shared back office silos, included no fewer than five representatives from the IT consultancy PA Consulting). The idea that information might be better organized locally, in doctors' surgeries or schools, has now had the backing of the leading think-tank on information policy issues in the UK, the Foundation for Information Policy Research. The big idea that vast software systems could encompass everything, and squeeze every possible purpose into their all-embracing protocols, seems slowly to be on the way out.

One of the most powerful voices calling for change is the digital musician and pioneer of virtual reality, Jaron Lanier, whose 2010 book *You Are Not a Gadget* describes the development of what he calls digital Maoism, where web-users become a new proletariat toiling for the benefit of an all-powerful virtual bourgeoisie. 'We're sending them to peasanthood, very much like the Maoists have,' he wrote.

Lanier's target has been the idea that individual creativity is being undermined by the internet, partly because so few people seem prepared to pay for it – I speak as a writer – and partly because the prophets of a digital future are toiling towards a day when there will be no individual books, pictures or musical compositions; just one 'mashed' whole. That was the idea behind Kevin Kelly's predictions in the *New York Times* in 2006, and it is the meaning behind the innovations known as Web 2.0.

Successful organizations are going to have to plough a different furrow. They are going to have to be able to tell the difference between creativity and second-rate mush. They are going to have to tell the difference between friendship and virtual networking. They will need to be able to see that, if their relationships with clients or customers get mediated though the categories that the software happens to allow, then they are going to be illusory and constrained. They will also need to see clearly where IT can hinder these relationships and where it can enable them.

'A real friendship ought to introduce each person to unexpected weirdness in the other,' says Lanier – and the same is true of any relationship between a company and organization and the people who use them. He is backed, rather reluctantly perhaps, by the founder of Sun Microsystems, Bill Joy: 'We can see glorious visions of stunning efficiency that truly could enrich humanity beyond our dreams. But we make a foolish and ancient error if we forget that quirky humans, who haven't evolved significantly in 20,000 years – and who still very much need interaction, recognition, and relationships – will have to make it happen and are supposed to be the beneficiaries.'

Find out more

Simon Head's *The New Ruthless Economy* is a key source for this chapter (Oxford University Press, New York, 2005). Simon Caulkin's columns have also tracked the phenomenon, and David Craig and Richard Brooks' *Plundering the Public Sector* (Constable and Robinson, London, 2006) provides a fascinating insight into the IT consultancy phenomenon. I would particularly look at Caulkin's column 'Good service must not follow the GM road to ruin' (*The Observer*, 7 June 2009).

The key re-engineering text is *Re-engineering the Corporation* (Michael Hammer and James Champy, HarperBusiness, New York, 1994). See also Sir David Varney's report *Service Transformation* (HMSO, London, 2006).

Which brings us to the IT scandals. One of the most revealing documents on the NHS computer project is the report by two MPs, Richard Bacon and John Pugh ('Information technology in the NHS: What next?', London, 2006, available at www.richardbacon.org.uk). The full story of the official mania for huge IT projects in the UK is in David Craig and Richard Brooks' *Plundering the Public Sector*. I also recommend Jill Kirby's pamphlet 'The reality gap' (Centre for Policy Studies, London, 2009).

Stafford Beer's essay is in Patrick Hutber's collection *What's Wrong with Britain?* (Sphere Books, London, 1978). Jaron Lanier's book is *I am Not a Gadget: A Manifesto* (Knopf, New York, 2010). See also www.jaronlanier.com.

Rule 8

Give everyone the chance to feel useful

Good institutions encourage self-assembly, re-use and repair. They do not just service people but create capabilities in people, support initiative rather than supplant it.

(Ivan Illich, *Deschooling Society*, 1971)

It has transformed the way we practice medicine. It has stopped us seeing our patients in terms of us and them, as if we were just service providers to people who are classed as 'needy'. We are no longer looking at them as bundles of need, but recognizing that they can contribute, and when you see people light up when you ask them to do so, it changes your relationship with them.

(Dr Abby Letcher, Lehigh Community Exchange, one of the most successful of the American time banks)

Summary

- When human skills get ignored or deliberately side-lined, then they atrophy.
- The human skills of *non*-professionals can have a particularly dramatic impact in a way that professional skills can't.
- Asking for help from people who have always received, and never been asked for anything back, can transform people's lives. It can also transform the organizations and make them human.

Julia Neuberger has had a fascinating career, as a rabbi, in the health service and more recently as Gordon Brown's volunteering tsar. But she provided a fascinating insight into why some institutions feel human and some don't, and – at least in the NHS – it doesn't have anything to do

with money. In her book about older people, she described with horror
how her uncle was neglected in three of the four hospitals in which he
lived his final weeks. She explained that the one exception was also the
hospital which was most cash-strapped:

> When my uncle eventually died, in the hospital which really under-
> stood and respected his needs and treated him like a human being,
> there were volunteers everywhere. In contrast, there was barely a
> volunteer to be seen in the hospital which treated him like an object,
> although it was very well-staffed. At a time when public services are
> becoming more technocratic, where the crucial relationships at the
> heart of their objective are increasingly discounted, volunteers can
> and do make all the difference.

Baroness Neuberger was writing shortly after some of the revelations of
just how badly old people were being treated in a series of hospitals
which had become particularly obsessive about government targets.
What she suggests is that volunteers are the antidote to this. In wards
where older patients might otherwise be mistreated or ignored, she says,
'the mere presence of older volunteers are the eyes and ears that we
need'. Human beings provide that kind of alchemy, however target-
driven the institution is around them.

It isn't quite clear why this is. Is it because the presence of outsiders is
a reminder to staff of what is important and how to behave? Is it because
it stops them getting too inward-looking, or prevents that brutal contempt
for customers that – as we have seen – can emerge in large organizations,
public and private? I don't know. But there is a new frontier opening up
in this debate about how human beings make things work, which
suggests that it is not just about having other people there. The volunteers have an effect because they are working alongside staff. It is
because the boundaries are blurring between the world inside the organization and the world outside. I don't know whether it would work the
same way if the volunteers were just there observing, but I suspect that
would just cause resentment. No, this is because they are equals. It isn't
just because outsiders are *watching* staff at work, it is because they are
sharing the work that it is so humanizing.

The trading company Trafigura that dumped toxic waste which
poisoned 30,000 Africans seemed unconcerned because the people were
so distant. 'Distance launders the bloodletting and technology purifies
it,' wrote the columnist Simon Jenkins in 2007, after listening to the
'ghoulish' conversation between American pilots in the Gulf War in

2003, accidentally attacking a British column. 'War becomes another video game,' he wrote. It is because the people who suffer are outsiders that they seem to matter less. In all those cases, a few elderly volunteers working alongside the Trafigura traders or American pilots might have done wonders for their sense of humanity.

But how can we blur the boundaries between inside and outside?

It so happens that blurring these boundaries is currently at the cutting edge of product development and marketing. Take two internet companies: Slim Devices and Threadless. Slim Devices sell a product called Squeezebox used to play music at home from MP3 players via a wireless connection. The Squeezebox was actually developed – not by in-house technicians – but by a network of enthusiasts and computer geeks around the world. Threadless customers go online to buy new T-shirts. That is the main idea, after all. But they can also put forward their own T-shirt designs, vote on other people's designs, and help the company refine what they sell next. 'In this world, running a company is not about brilliance or command,' said *Fast Company* magazine in 2007, 'but about attracting and orchestrating the work of talented and passionate outsiders – people who know more than you do, have better ideas, and maybe even care more about your product than you do.'

What this means is that, although you still win in business by having the best employees – as you always did – you also win by having the best enthusiasts as customers. Sometimes you win spectacularly. Slim Devices took in $10 million in 2006. Its founder Sean Adams refused to apply for patents for the devices his customers developed, because he said he couldn't sleep at night if he did. As always, when it comes to using human beings to their maximum, this is about letting go of the myth of absolute control. 'Do I make decisions myself about changing the product, or do I open it up?' Slim Devices head of technology Dean Blackketter told *Fast Company*. 'Every time I've opened it up, it's paid off.'

The Indian management thinker C. K. Prahalad called this process 'co-creation'. He included in this Facebook, where the customers also fill in the blank pages that are provided by the company. Or iPod, which Apple created as a blank canvas which customers can fill with their musical tastes. Not to mention Zagat, eBay or Skype, all of which are new spaces that people can populate as they want. But he also includes companies that are demanding something back from customers too. Such as ICICI Prudential in India, which is offering lower health insurance premiums to diabetes customers who exercise and meditate.

The most famous application of this idea is an online encyclopaedia. Larry Sanger and Jimmy Wales dreamed up an online encyclopaedia called Nupedia, and signed up leading academics to write it. But the process was incredibly slow. They received only 24 entries in two years. 'The pace was horrible,' said Wales. Early in 2001, they discovered the idea of wikis, which anyone could edit, and re-thought the whole project. The academic editors didn't like the idea at all, but the result is now one of the most consulted sources of knowledge in the world. 'When I launched Wikipedia, we surpassed the entire output of Nupedia within a matter of days,' says Wales.

Prahalad and his colleague Venkat Ramaswamy coined the term 'co-creation' in an article in the *Harvard Business Review* in 2000 entitled 'Co-opting customer competence'. They said that corporate value was something which staff and customers created together, but they said that it meant more than just developing products – it meant people interacting with their favourite company in a whole new range of ways. It meant companies such as Nike giving customers the tools to design their own sneakers, or Lego letting people download software for designing toys which they then manufacture.

But the real cutting edge is companies such as Cisco or Apple trying to involve customers in a range of other ways, taking advice from customers who really push products or services to extremes, or from academics who are developing their own passions in ways that can also develop the company. Companies such as Nokia and IBM are trying to take this further to look at how they can involve customers in developing training or performance management. This is the brave new world of corporate thinking: blurred boundaries.

Blurred boundaries are not entirely new. There has been a systematic blurring of the boundaries between supermarkets and their suppliers, sharing the same systems, or between retailers and their franchisees. Letting the general public in is a new idea in business, and it is bound to humanize, as long as the companies involved never take that input for granted. But there is a similar process, just like co-creation, which is happening in the public sector, which is not subject to theories from the world's business gurus, and it is happening in schools, hospitals and courts on both sides of the Atlantic – and it is all about providing a whole new way of making organizations more human, and therefore more effective.

When Professor Elinor Ostrom won the Nobel prize for economics in 2009, it was doubly unusual. No woman had ever won it before, and she was not even an economist. She is a social scientist whose understanding about what makes organizations work effectively is directly relevant to

Julia Neuberger's insight about her uncle's hospital. She had been at Indiana University for most of her career, spending much of her time in the 1960s, by her own admission, getting her own students out of jail after anti-Vietnam War protests. It was the period which saw the huge consolidation of schools in the USA, reducing the number of school districts from 110,000 to just 15,000 in the four decades up to 1950. There was a similar consolidation of hospitals, police forces and companies. This was the emerging era of big corporations that, as we saw in Rule 4, have tended to constrict the role that people can play in making them effective.

Ostrom was fascinated by the issue of scale. When she scraped up enough money to begin research, she decided she would look at the same phenomenon that was happening across the USA of consolidating police forces into larger and larger units. She had just enough money to hire cars for her research students to drive around Indianapolis for ten days, and test out the effectiveness of different styles of policing. It was pretty clear, after the results came in, that the smaller the police force, the better they were at responding to emergency calls.

This challenged conventional thinking, which assumed then – as it still does – that bigger is better. Her black students urged her to have a proper look at policing in Chicago, and sure enough, it was the same there. The small police forces in the black suburbs were just as effective as the huge police force covering central Chicago, which had 14 times the funding. The Chicago police were interested in her research and asked her why she thought the crime rate seemed to be rising when police shifted from walking the beat to driving round in patrol cars. It was here that she made her real breakthrough, understanding how much the police need the public, and how cut off they had become.

She needed a word that described that indefinable cooperation between police and public which was so easy when they walked around. She called it 'co-production'. If the police forget how much they need the public – and disappear into technocratic systems, patrol cars or bureaucracy – then crime goes up. It is the same pattern with doctors: they need the cooperation of patients if they are going to make them well.

The argument about what causes crime is one of those completely unanswerable and unresolvable arguments that will go on and on. If you are on the Left, you tend to think it's about poverty; if you are on the Right, you tend to think it's about how people behave when they can get away with it. In fact, one of the biggest studies which set out to answer the question found, rather to the surprise of the researchers taking part, that it was actually something different. Once again, it all happened in Chicago.

The project dates back to 1995, two decades after Elinor Ostrom began her research there, when Chicago sociology professor Robert Sampson led a team from the Harvard School of Public Health. They were determined to pin down what made the difference between high crime areas and low crime areas, especially when it involved violent crime. Ever since Al Capone and the St Valentine's Day Massacre, Chicago has had a reputation for spectacular criminality, but then Sampson was based there, so it made sense to concentrate on his home territory. What he and his team did was to divide the Windy City into 350 different neighbourhoods and devise an exhaustive questionnaire to find out what each area was like. There were nearly 9000 interviews. Then they compared each profile with the rates of violent crime.

At first nothing made sense. None of the usual measures – poverty, racial discrimination – seemed to have much to do with the crime rates. But there was indeed a difference between the risky places and the safe ones, and they called the critical factor 'collective efficacy' – which meant the willingness of local people to intervene with children when they saw them playing truant, painting graffiti or hanging threateningly around in gangs on street corners. It was what the team called a 'shared willingness of residents to intervene and social trust, a sense of engagement and ownership of public space'.

Sampson went on to confront the prevailing idea – the Broken Windows theory – that rubbish and small signs of disorder led to more serious crime. He and his team rigged up a video camera in the back of a car and drove around Chicago for nearly 250 miles. Once again, the received wisdom didn't stand up: there was graffiti and rubbish in both the safe and the risky areas. Once again, it was Sampson's 'collective efficacy' that made the difference.

The Chicago crime study was important because it revealed something that people had suspected for some time, but had never quite been able to pin down. Research already showed that people were healthier when they had friends. It was known that doing business was easier for people if they lived in a network of people. But violence? That was just something about poverty, wasn't it? The Chicago study showed that it wasn't – or at least, not directly. It was to do with whether people trusted each other or not. If they didn't, the crime rate was high; if they did, that bizarre alchemy that human beings bring to a situation seemed to transform a place no matter how poor it was.

Sampson has recently moved to Harvard and developed the idea in a radical new direction. He believes the reason the crime rate has gone down right across the USA in the last generation is because more immigrants have come in. Mexican and Latino communities are better at this collective efficacy, because they tend to have a stronger family culture. They deal with the risk in their neighbourhoods much more informally. This may be an intellectual leap too far, but Sampson's basic idea changes a great deal about what we believe about public services. It means they have to create the conditions for their own success. It isn't enough for them just to fix people's problems one by one – that way leads to exhaustion – they also have to reach out and deal with the causes.

If one of the causes of high crime or bad health is that people don't trust each other, then they have to reach out somehow and stick the neighbourhood back together again. They have to ask ordinary people for help. Just like Wikipedia or Linux, police forces or health systems need the help of their customers, they need to blur their professional boundaries.

I remember watching a friend of mine wander through one of the bleakest housing estates in south London, where she lived. For anyone who didn't know the place, it was a little nerve-wracking. But as we went, she greeted some of the children and young people by name, told some of them off, hailed others, laughed with some others. I realized as we went that this was precisely the 'collective efficacy' that Sampson's team talked about. It wasn't delivered by a government programme. It wasn't even measurable. It was the kind of thing that a super-catalyst can achieve just by wandering regularly through somewhere. Management By Walking Around again.

But how do you use this idea to make a difference? When Sampson's study was first published in 1997, it was picked up immediately by one of America's most innovative human rights lawyers. Edgar Cahn was then ten years into his experiment with a new system he called time dollars or time banks, which paid credits to people for helping out in their neighbourhood. When they helped someone, they earned time; and when they needed help themselves, they spent time. Cahn's time banks would lead to a whole new way that organizations could get their customers to deliver services.

Cahn came up with the idea in hospital after a heart attack in 1980 because he hated feeling useless. He had no sense that all that the money he had spent on health insurance was finally going toward something worthwhile. Quite the opposite. His whole life had been dedicated to being useful. He was the son of an eminent professor of jurisprudence and had been one of the bright young backroom boys of the Kennedy presidency, writing speeches for Bobby Kennedy as Attorney-General. That had led him, in turn, to a key role in President Johnson's War on

Poverty. He had also set up the Antioch School of Law to inject a different kind of on-the-job legal training into the system, alongside the elite law schools of Washington, DC. It was defending Antioch against closure which had given Cahn his heart attack.

'My idea of myself was someone who could be special for others, who could do something they needed,' he said. 'And here I was, a passive recipient of everyone else's help. I didn't like it. There was nothing I could do. Or was there? The question wouldn't go away. It never has. It became a very personal fight. I refused to be one more throw-away person. And I knew that the fight was not just about me.'

There was no going back. He went into hospital a conventionally progressive 'liberal' lawyer, who had cut his teeth in the civil rights movement. He came out something completely different: a critic of conventional welfare, a radical who would shortly be taking up his cudgels against over-bearing professions, but – most of all – determined to find a better way of using people's human skills, rather than allowing them to go to waste simply because they weren't marketable.

The time bank idea was a whole new way of rewarding people for doing those critical but unpaid jobs in the community that desperately need doing, from tutoring in schools to visiting old people, and which the market doesn't normally reward. He described the idea like a blood bank or babysitting club:

> Help a neighbour and then, when you need it, a neighbour – most likely a different one – will help you. The system is based on equality: one hour of help means one time dollar, whether the task is grocery shopping or making out a tax return … Credits are kept in individual accounts in a 'bank' on a personal computer. Credits and debits are tallied regularly. Some banks provide monthly balance statements, recording the flow of good deeds.

The idea was taken up first in 1987 by a health insurance company in Brooklyn called Elderplan. Their Member to Member scheme was supposed to use the efforts of their younger customers to help keep their older customers living at home. That worked, but the real impact was unexpected. The biggest health impact wasn't on the older people who were being helped, it was on the slightly younger people who were volunteering.

The result was another kind of blurring of boundaries. Elderplan was an insurance company that was suddenly organizing a transformative mutual volunteering network of more than 10,000 people. There were walking clubs, phone quizzes and befriending schemes, and a great deal else. People could even pay a quarter of their insurance premiums using the time they had earned. To Elderplan's surprise, the scheme became a selling point that differentiated their service from their rivals. When their DIY team launched, with volunteers doing small repairs for other members in return for time, they put it on their posters. One carried a picture of a DIY team member, complete with hat and spanner, with the slogan 'Does Medicare send you a friend like George?'

Another one asked: 'Does Medicare lift your spirits?' When the American healthcare industry was being accused of cynicism because of its apparent inhumanity, Member to Member was demonstrating a very powerful human alternative. There are now similar organizations in healthcare settings and in schools and housing estates in the UK too. Some of the most ambitious are in doctors' surgeries. They increase in complexity, person by person and relationship by relationship, and they absolutely refuse to be categorized or to specialize. If befriending someone, or fitting new light bulbs in their home, is going make someone healthier, then the members of the time bank make it happen. They are not part of the medical profession, but they are delivering medical services.

When Edgar Cahn came across the Chicago study about crime, it provided him with the evidence he needed to explain why time banks were effective. They were not just blurring the boundaries between professionals and clients, or between inside and outside the insurance companies or health centres. They were also stitching together the local neighbourhood so that it could insulate itself against crime and keep people healthy. Elinor Ostrom was using the term 'co-production' to explain what was wrong with institutions like the police, but Cahn began using it to explain what was wrong with the economy, when it deliberately de-valued people's human skills and forced them to work in McDonald's rather than making families and neighbourhoods work better.

'We are strip-mining our communities,' wrote Cahn in his 2000 book *No More Throw-away People*:

> We are depopulating our neighbourhoods, we are atomizing families. The process keeps accelerating as every asset we can lay our hands on gets sent to the market for sale at the prevailing price. The market is governed by a pricing system that devalues precisely those activities most critically needed in communities … The trouble is that the vast bureaucracies, the armies of professionals, the complex systems of targets that we are throwing at the problem are not just making the situation worse – undermining the ability of people and families to

use the assets they have – but they are tackling the warning signs, the statistics, without trying to staunch the basic problem.

Cahn used the analogy of a computer, which runs powerful specialized programmes, all of which rely on a basic operating system without which they can't work by themselves. In the same way, our specialized services dealing with crime, health or education, rely on an underpinning operating system that consists of family, neighbourhood, community and civil society, bringing up children, looking after old people. He called this operating system the 'core economy', borrowing the phrase from the economist Neva Goodwin. It is vast, in both its range of activities and its economic impact, but it doesn't rely on price to enable it to be exchanged.

How could it? For prices to be high, things have to be reasonably scarce, but the attributes of the core economy are absolutely abundant. They are human abilities to look after each other, to empathize, bring up children, tell stories and they are absolutely everywhere – so the money economy assumes they are worthless, allows no time for them, devalues them and pretends they don't exist. So much so that eventually they start to disappear. Ostrom argued that co-production was vital for organizations; Cahn argued that it was vital for society.

It was a different aspect of how organizations need human beings if they are going to work effectively. We have seen how they blind themselves to the skills their employees have. Cahn and Ostrom were talking about how blind they are to the skills their customers have too, and their customers' neighbours. When these human skills get ignored or deliberately side-lined, then they atrophy.

But Cahn went even further. If the demands on doctors or police just carry on growing, it isn't because public services have failed to consult people enough, or even that welfare services have paid people too much – or any of the conventional explanations. It is because they failed to ask people for their help and to use the skills they have.

He argued that this is the forgotten engine of change that makes the difference between systems working and failing. Instead, people's needs become the only asset they have. It isn't really surprising, in that case, that their needs grow. Some people try to stop getting better, like the mother who kept her son in a wheelchair and fed him from a tube for five years – though he no longer had symptoms – in order to keep the benefits flowing. They burnish their depression or ME because they fear that nobody will help them any more if they get better. Of course they do: it's the only way they can get any attention.

It was Cahn who wrought one of the strangest turnarounds in the history of Chicago's troubled public school system and, if you can do it there, you can probably do it anywhere. The system had always been troubled, even before the campaign a century ago to limit class sizes below 70 pupils. Even now, 80 per cent of pupils are low income and the drop-out rate is high. By the 1990s, it was ranked as bad as the school system in desperately dysfunctional Washington, DC, at which point Mayor Richard Daley thought enough was enough, and he appointed his chief budget officer Paul Vallas to sort it out. Vallas decided the solution was to recruit 10,000 volunteer tutors and to flood the system with them.

This was a revolutionary plan in itself, but it had one major flaw. The vast majority of those tutors were professionals. They were often former teachers themselves, or sometimes they were the brightest pupils brought in as volunteer peer tutors after school. It was their skills as semi-professionals that were being recruited, and often they had little or nothing in common with some of the children they were teaching, especially from notorious 'killing zones' such as Englewood, where the whole idea of learning anything was a source of deep suspicion. Most of the schools offered their brightest children to take part and the local education company that won the main contract took them on and taught them how to tutor.

The problem was that disaffected young people don't see the world quite as professionals do. In fact, the research suggests some of them don't learn because they are afraid their friends will make fun of them, especially in failing schools. In some places, it is actually dangerous to do anything that might seem like you are looking for approval from the teacher. Let's face it, it certainly isn't cool. So when Edgar Cahn met Vallas in 1995, he urged him to do something seriously revolutionary – to recruit tutors from the *least* successful schools.

He set up shop in five elementary schools in Englewood the following year, and the next year there were ten, run by a telecommunications engineer called Calvin Pearce. Nearly 200 children were involved in each school, and earned a recycled computer to recognize what they had done. When Cahn came back and watched the peer tutors at work, he asked them afterwards what they had learned. One said: 'I learned that when my kid did her homework well, I should stick a label on the paper and write "you are a smart kid", so that she could take it home and show it to her mother.'

Another one said, rather elusively: 'I learned that there are words inside of words.'

Pearce decided that, in his schools, they would accept any child who volunteered to be a tutor, whoever they were. It was a risk, and the education authorities – desperate as they were – looked on askance, but it turned out to be the best decision they made. Many of those coming forward were so-called problem children. Some of them were being treated for attention deficit disorder or had special educational needs. But they had what the bright children didn't have: a conviction that, if *they* could do a maths problem, then anybody could. That had the opposite effect of what you might imagine. They had very high expectations for the children they tutored.

It soon became clear that the peer tutoring was a great success. Truancy was down on the days it was taking place after school. But there were a number of unexpected effects too. The tutors were having an impact, not just on their pupils, but on themselves. Their own work was getting better and, at the same time, bullying was going down. There was less fighting after school because the tutors took it upon themselves to protect their pupils. Nobody asked them to, they just did.

One mother who dreaded going into the school, because she was only ever asked to go there to get appalling reports about her child's behaviour, suddenly found she was proud to go. Her son was a tutor, and not long afterwards, she was volunteering to help in the school too. More than ten years on, the peer tutoring programme has helped more than 4000 pupils in 45 schools, mainly on the South Side of Chicago, and it had spread to other American cities.

There was nothing revolutionary about peer tutoring in itself, which emerged in the USA in the 1990s. What was most important about the programme was that it wasn't the semi-academic skills of the brightest pupils they were using, or the professional skills of the adult volunteers. It was the human skills of the very ordinary pupils instead, and – most successful of all – they managed to turn cool 15-year-olds into advocates for learning among their slightly younger peers.

What the peer tutoring programme showed in Chicago was the power of ordinary human skills: a bunch of largely untrained teenagers, many of them considered problem children by the school system, were able to bring about the kind of changes that had stymied hugely expensive government programmes involving coachloads of trained teachers and evaluators. But the Chicago success implied something else. The human skills of *non*-professionals can have a particularly dramatic impact in a way that professional skills can't.

If blurred boundaries are the future – bringing in human skills to make things happen – then most organizations are organized in precisely the wrong way. Most of the public sector is still delivering set services to grateful, passive recipients. In the same way, most businesses are structured in exactly the opposite way that they need to be to get the best from their customers. They are designed to deliver services that they own and dominate to grateful consumers. They have hardly any of the skills they need to inspire customers and set them to work as part of a joint endeavour.

Both public and private sectors could theoretically learn from the voluntary sector, except that it is going in the same mistaken direction, merging into vast, risk-averse, over-professionalized monopolies, and subject to the targets and controlling IT systems that tend to side-line human skills the most. The next revolution in public services seems likely to lie here – to shift the way services work, in schools, hospitals or justice, to involve the people they are working to help as equal partners in the delivery of the service. The evidence is that asking for help from people who have always received, and never been asked for anything back, can transform people's lives. It can also transform the organizations and make them human.

Find out more

Edgar Cahn, the originator of time banks and the pioneering co-production theorist, is the key influence in this chapter, and his book *No More Throw-away People* (Essential Books, Washington, DC, 2004) is a vital source. So is his website www.timedollar.org. There is information about UK time banks at www.timebanking.org and about the time banks I mention here at www.rgtb.org.uk and www.pgtimebank.org. See also Cahn's pamphlet 'Priceless money' at http://coreeconomy.com. The Lehigh Community Exchange research was published as *Building Community Ties and Individual Well-Being: A Case Study of the Community Exchange Organization* (Judith Lasker and others, Lehigh University, 2006) is available at is at www.lehigh.edu.

Cahn's website includes an important section on co-production. Some of Elinor Ostrom's key texts include her original book *Community Control and Government Organization* (with P. Whitaker, University of Chicago Press, Chicago, 1973). Also her *Community Organization and the Provision of Police Services* (Sage Publications, Beverly Hills, 1973) and 'Crossing the great divide: Co-production, synergy, and development' (*World Development*, vol 24, no 6, June 1996). See also Elinor Ostrom and Gordon P. Whitaker, *Community Control and Government Organization* (University of Chicago Press, Chicago, 1973).

Anna Coote was an important pioneer of the idea in the NHS (see her new pamphlet 'Green Wellfair: Three economics for social justice', New

Economics Foundation, London, 2009). Probably the easiest introduction is *Co-production: A Manifesto for Growing the Core Economy* (Lucie Stephens, Josh Ryan-Collins and David Boyle, New Economics Foundation, London, 2008). Also useful is *Aspects of Co-production* (David Boyle, Sherry Clarke and Sarah Burns, New Economics Foundation, London, 2006) and the final report of the same project (*Hidden Work*, David Boyle, Sherry Clark and Sarah Burns, Joseph Rowntree Foundation, York, 2006). You will also find more information about co-production on my own website at www.david-boyle.co.uk/systems.

More recently I have co-authored a series of three reports with my colleagues at the New Economics Foundation and NESTA, *The Challenge of Co-production* (2009), *Public Services Inside Out* (2010) and *Right Here, Right Now* (2010) which are all available on the co-production website at www.coproductionnetwork.com. See also the academic network www.coprodnet.org.

The Chicago crime study I refer to is by R. Sampson, S. Raudenbush and S. Earl ('Neighbourhoods and violent crime', *Science*, 15 August 1997). Professor Sampson's later research can be found as 'Rethinking crime and immigration' (*Contexts*, Winter 2008). The story about Julia Neuberger's uncle is from her excellent and humane book about ageing *Not Dead Yet* (HarperCollins, London, 2008). The article about corporate applications of this idea is by Alan Deutschman, 'Ears wide open', in *Fast Company*, December 2006.

Rule 9

Make organizations into engines of regeneration

Trust transparent types. Oppose opaque operators. Trust experts who make you feel smart.

(Francis Gouillart, business co-creation pioneer)

Charity wounds.

(Mary Douglas, in her introduction to Marcel Mauss' *The Gift*)

Summary

- When people are just given things and are never asked for anything back, there is a subtle corrosion of self-esteem, but two-way relationships seem to make clubs and societies 'sticky'.
- Reciprocal links have to go both ways. Sticky organizations expect people to be both givers and receivers, and they reach out into the surrounding neighbourhood to build these complex links.
- Blurring the boundaries between inside and outside needs to create at least some equality in the relationship. If it doesn't do that, then getting your customers to help you save money is just manipulative.

There was a time when most neighbourhoods, and the poorer ones especially, were alive with choirs, reading clubs, friendly societies, pigeon breeder clubs and all the rest. Then something happened. They managed to get purpose-built community centres and attracted grants for permanent professional staff to manage them. The old charities that used to be run by small groups of friends in churches or working men's clubs either died out or were pushed aside by the new lottery-funded generation. Now the community centres in the Welsh Valleys, for example, usually have full-time managers and staff. The peculiar thing is that they are often almost empty. What was once a network of voluntary projects and chapels gave way to a series of professional agencies, with paid staff, delivering services to passive consumers. Where did all that energy go?

In some ways, this is just an aspect of what the American sociologist Robert Putnam called 'Bowling Alone', a terrifying description of the erosion of neighbourhoods – watching the gentlemen of New London, Connecticut reduced to sitting alone in the local bowling alley, staring sadly upwards at the television.

But there is something else going on here, and I chose the Welsh Valleys as an example deliberately, because that is where a youth worker called Becky Booth began to ask difficult questions about the local community centres, many of them state-of-the-art buildings paid for out of regeneration funds following the closure of the local coal mines. She was working as part of a team at the University of Wales that was looking at setting up time banks in the Valleys, but – time and again – she came across the same phenomenon.

'I kept on finding these empty community centres,' she says now. 'They were often fantastic buildings, with loads of resources but hardly any people using them, or they were run by small groups of people who controlled who could come in, while the local young people hung about outside. I would go in and there would be nobody there. It was such a waste.'

Often when the teenagers did come in, they came in for a specific youth event at a set time and then went again. In fact, it was the mismatch between local young people who needed somewhere to go, and well-funded but empty community centres, that first made her realize what was wrong. She had been in the Far East writing a thesis about how international charity had undermined social networks in the areas hit so badly by the tsunami in 2004. She realized she was seeing the same phenomenon here.

'It seemed obvious that they had no sense of ownership over it,' said Booth. 'The community centres had begun when there was a whole network of people, in chapels and miners' institutes, who were teaching people to read, making things happen, and paying a contribution to do so, when there was a natural sense of give and take. Now it was all about coming in and just receiving – so they had no relationship with the centre.'

She went on to launch a series of projects in the Valleys that have played a role in involving people in a more active way. One of these is run by Taff Housing, a housing association with more than 1000 homes in some of Cardiff's most disadvantaged housing estates, where tenants earn credits by volunteering their time to help deliver the services of the housing association – often just being good neighbours. They can spend these credits through a partnership that Taff have negotiated with Cardiff's leisure services, Cardiff Blues rugby club and the Gate Arts and Community Centre, which accept credits earned by tenants instead of cash to use their services.

Where this happens, it can mean a huge shift of focus for public services. They are no longer looking inwards so obsessively towards targets and procedures, but they find themselves looking increasingly outwards at the local neighbourhood to create supportive social networks, seeking out local energy wherever it is to help deliver services, making their services broader and more human but also seeing clients for what they *can* do, not just for what they need.

Behind all this is the strange untold story of community development, especially when lottery funding is involved. When local agencies or charities discover a local need, they apply for grants to tackle it, most of which goes on their salaries. Then the mystery: as people discover the service, the need seems to grow. Agencies have to ration support. Then the grant runs out and everything has to be applied for again to keep people in jobs, but dressed up as something wholly new and innovative. The handful of local people who had been genuinely involved get dispirited, and the agency starts looking around for another need that could be packaged as a grant application.

It's easy to be cynical about this. One academic has called it 'farming the poor' and it is true that charities need poor people at least as much as poor people need charities. But there is a real problem: the takeover of genuine local action by professionals has undermined those links of obligation and mutuality that made things happen and that actually – in the end – may make life feel worthwhile. As a result, we have increasingly empty local institutions, empty community centres and a substructure of busy charity professionals who have failed to rebuild the sense of mutual support on which everything else depends. They haven't learned the lessons of Elinor Ostrom and Edgar Cahn.

Cahn came up against this problem quite by accident when he was defending his National Legal Services Programme, the service that helped organizations to sue the government to enforce their rights. He had urged the programme over the years to ask the people they were helping to give something back, but they never quite got round to doing so. Then suddenly, in 1994, there was a Republican landslide, determined to reduce the federal budget deficit, and a young maverick called Newt Gingrich was in the House of Representatives, looking for ways of saving money. The Republicans had never much liked the Legal Services Programme anyway, so the scene was set for the inevitable congressional hearings before it was shut down.

'Key strategists thought that the usual mobilization of bar support would save it,' says Cahn. 'It would be tough, they said, but it was just a matter of digging in and working hard and the usual fight would end the usual way. My crystal ball read the picture differently: I tried to persuade them to enlist the clients of legal service programmes as 'co-producers'. Clients would pay for their help in time dollars, paid off helping out in the community for passing on what they have learned.'

This meant that, for the first time, clients would be paying back. They would also be increasingly involved in self-help, but – or so Cahn reasoned – there would be another side-effect too. There would be people all over the country who were more than just grateful, passive recipients, and a constituency that could defend all these new organizations when their grants were threatened. But, in the end, only the California Rural Legal Assistance programme agreed. Cahn wrote to the head of the National Legal Aid & Defender Association. There was no answer. He published an article in the prestigious *Yale Law Journal* explaining what he wanted to do. Silence.

The programme was duly cut by a third and hamstrung in other ways. The hearings were held in Congress. But out of the three million people a year that the programme had helped for 33 years – that's about 100 million households – not one client turned up at the hearings to defend it.

A year or so later, Cahn's own law school was also under threat. This was the successor to Antioch, the District of Columbia School of Law, which was modelled on a teaching hospital. Students follow the proposals that Cahn put forward. They go out into the community and give legal help, but they don't just *give* it. They ask for something back through one of the time banks in a Baptist church or in local housing complexes. It was a difficult campaign to win, given that Washington already had six law schools and a massive budget deficit. Even *The Washington Post* was calling for it to be closed. But hearings organized by the District of Columbia Council didn't go the same way as the ones in Congress. Those who had been helped, and paid back, came out in droves to support the law school and it stayed open. Giving something back for the help they had received had made people defend the law school. Perhaps because it was more equal. It wasn't charity any more.

Cahn describes this as the power of 'reciprocity'. It provides an accelerating energy for organizations and businesses. Without those ties of obligation, and that sense of a relationship, the energy dips and dissipates;

with it, the energy carries on. When there are these reciprocal links between people, the organizations seem to get a kind of stickiness about them: they generate a power to keep people involved. The mutual building societies were theoretically owned by their customers, but there were no real links with them – people didn't have to *do* anything to be members. Perhaps that explains why so few people defended the building societies, and they were sold off and swallowed up by the big banks. On the other hand, companies involved in this kind of co-creation do seem to be building genuine loyalty. And it sticks.

Something else is going on here that makes these successful organizations 'sticky', which keeps people's loyalty and engages their imagination and energy. That something seems to be an antidote to the inflexibility of one-way relationships. When people are just given things, and are never asked for anything back, there is a subtle corrosion of self-esteem. There is no relationship to keep things going except a rather corrosive dependence. In short, as the anthropologist Mary Douglas put it, 'charity wounds'.

The next step in the story of sticky organizations does come from anthropology. Another anthropologist, Polly Wiessner, originally went to the deserts of southern Africa as an archaeologist, but she didn't enjoy it. She became fascinated instead with the way the !Kung bushmen in Africa, where she was staying, dealt with risks such as famine or war. Now attached to the University of Utah, she has studied in Germany, Vietnam and Papua New Guinea, but she always comes back to this original insight: the amazing networks of reciprocal obligation that she found there, not just with each other – but with families hundreds of miles away.

These distant relationships are absolutely vital if the whole area is devastated. She was in the Kalahari Desert in 1974 when torrential rains destroyed the harvest, and watched while, one by one, the families made the trek to stay with their friends where there was enough food. The links with these distant families might have been inherited for generations, but they were there in an emergency. Thanks to new game sanctuaries, poachers and arbitrary official lines on maps, these distant partners are dwindling for the bushmen. But some links carry on and any extra food or resources they have is still given away to facilitate these long-distance ties of obligation.

It means in practice that people in the Kalahari talk about what they owe their friends – the distant ones and their neighbours – the whole time, and whether they are really in need when they ask for help. It is a tireless and exhausting subject of conversation. 'But this is what makes human beings unique,' says Wiessner. 'We are the only species that maintains social relationships of reciprocity. Being embedded in reciprocal systems like this is wonderfully secure. But it is a double-edged sword. It is actually rather a burden, because you are constantly being asked for things.'

She goes back there nearly every year, and she always finds the same thing when she comes home to her university. At first, there is a huge sense of relief to have escaped these intricate networks of obligation. Then, five days later, she suddenly feels a deep sense of loss and loneliness. Over the years, she has come to believe this is because these reciprocal ties of obligation are part of being human. 'It means people are constantly watching you, and you are never alone,' she says. 'There are a whole range of costs. But there is a trade-off between being needed like this and the sense of lack of purpose to life without it.'

This revelation led her to look more closely at reciprocity, and why people need it. The great sociologist Marcel Mauss said that people use giving and receiving as a way they build relationships. Children learn how to do it from about eight months old. When the Canadian psychologist Jim Rilling and his research team took brain scans of people while they were playing the Prisoner's Dilemma – the classic researcher's game where people have to cooperate in order to win – he found that the brain lit up in the pleasure areas when they trusted people enough to give and receive from them. Reciprocity has psychological rewards too. We are hard-wired for it.

Ever since Jack Kerouac hit the road, we have had a bit of a problem with the idea of community ties. It is true that, when people leave home – and taste the freedom from the tyranny of knowing absolutely everyone you run into in the street – it is hard to go back. People often need to escape into the anonymity of the adult world, where there are none of these exhausting ties of mutual obligation. The trouble is that we never quite grow out of this in modern culture. We are vigilant against entangling ourselves in anything likely to add to the demands on us later.

I remember years ago when an elderly man fainted in front of my car, I picked him up and took him where he wanted to go – the local Conservative Club, in fact. I helped him to a seat and explained to the manager what had happened, and suggested he keep an eye out. He said nothing, but his wife did. 'You've done a terrible thing,' she said to me, indicating her husband. 'You've given him responsibility.' This absence of mutual responsibilities is a kind of adolescent dream, encouraged by the prevailing culture, and – heavens – I share in it myself. But, as Polly

Wiessner's experience with the !Kung bushmen showed, it isn't very human. We find these ties irritating sometimes, and long to escape them but, when we actually do, we feel bereft.

They also have a secondary effect: they keep people active, which is also probably why they sometimes feel like such a burden. They make clubs and societies and other projects *sticky.* People feel the obligations to each other and, generally speaking, they stick at it. They don't stick at it if there is something obsessive about the other side, whether it is individuals or organizations. Or if the organizations feel inhuman, or too mechanical at their heart. They won't want to be involved in helping a company if its products are made by semi-slaves in Far Eastern sweatshops. But if those exceptions don't apply, and there is enough of a reciprocal exchange to keep the organization sticky, then communities begin to cluster around it, as they have in Member to Member, the Lehigh Community Exchange, Wikipedia or Lego. Reciprocal links have to go both ways. That means no more giver and receiver. Sticky organizations expect people to be both, and they reach out into the surrounding neighbourhood to build these complex links.

This is a new prescription for organizations. They don't just have to involve their customers in the delivery of services, especially public services – and especially those who have never been asked to give anything back before. They have to build reciprocal links into those relationships.

That is difficult for old-fashioned organizations to do. Making reciprocal relationships makes them nervous. It isn't what doctors or architects are trained to do. But it is slowly beginning to happen, from Edgar Cahn's peer-tutoring projects and the time banks in doctor's surgeries to cooperative nurseries run by the parents or cooperative supermarkets run by the customers – or courts where the judges are neighbours. There is a refreshing equality about these projects, but they also have that elusive stickiness – people stay involved and they defend the service if it is under some kind of bureaucratic assault. It means not just giving advice on services, or sitting on decision-making committees, but actually doing the work. This is what Cahn means by 'co-production' and it is about *doing* things. Not the same things, of course. Not brain surgery. But the kind of outreach and mutual support that ordinary people can actually do better than professionals.

One of the difficulties for the pioneers of this idea is that they often seem to fly in the face of the way public services have been developing over the past generation. They rely on face-to-face influence when the trend has been virtual. They appeal to general skills when the trend has been increasingly specialist. They believe in ordinary skills, amateur in the best sense, when the trend has been increasingly over-profession-alized. Most important perhaps, this kind of stickiness relies on the idea that the users of services, and their families and neighbours, are a vast untapped resource – when the trend has been to regard them as drains on an overstretched system.

Sticky organizations are different. In practice, they mean another nail in the coffin of the idea that all organizations are just glorified factories. They mean a huge and unprecedented mobilization of unpaid involvement by public service users, their families and their neighbours. It means a massive increase, not so much in volunteering because it will be outside the conventional volunteering infrastructure, but of mutual support and activity organized through the public sector. It means that every school, surgery, hospital or housing estate would become – as its fundamental purpose – a hub of increasing local activity.

It also means that some visits to the doctor will involve questions about what you can do to help, not just about your symptoms and matching drugs, but where your passions lie and what your objectives are in life. This is not something, of course, that is going to transform every fleeting brush with the NHS, but – given that 80 per cent of its time and resources are devoted to struggling with chronic problems – it will mean a major shift in direction. It means that something is asked for by professionals in return for the support they give. It isn't made into a condition – they will be given help anyway – but they will be asked to give something back as part of the continuing support or treatment.

This does not mean an end to state services. Quite the reverse. But it means an end to the language of 'services', and a re-organization of those services as nodes of multiple networks, as social catalysts which can reach out into the surrounding neighbourhood with a specific objective of prevention. It will also mean an effective meshing of different kinds of service, often on the same site – schools as clinics, housing estates as courts – which will make them much more personal and much more local.

Companies such as Cisco or Lego have flirted with co-production, but this is something different and more revolutionary. They are about to be overtaken by a new kind of public sector, with complex relationships rather than complex metrics at its heart. It is one that will require huge training and probably major investment to achieve in the long term. It will mean different kinds of skills, different kinds of buildings and different kinds of systems. That investment will be justified because of the revolution in effectiveness that it will bring. But it can also begin without that investment and immediately, at any level, because the basic resources are already in place. They are huge and largely untapped: the armies of passive clients of public services, their families and neighbours. We are living through a period when we believe resources are terribly scarce, but these are not scarce at all.

Some of the most dramatic new projects which put these resources to use are having a huge effect, like Cahn's innovative Youth Court in Washington, DC. There is something wrong with the business of law and justice in Washington, when you think that – despite all those law schools – half the black men under 24 are in prison, on remand or on probation. Given that the youth justice system there is failing so spectacularly, it seemed like a good idea to extend the model of sticky organizations to youth justice, which is how Cahn came to run part of Washington's youth justice service, organizing his own courts for young people arrested for the first time for non-violent offences, who were then tried before juries of other teenagers.

The jurors are paid in credits for being there, which they could use to buy a refurbished computer, like the Chicago peer tutors. Their sentences might involve community service or writing letters of apology, but also mean they have to sit on a jury themselves later on. The juries are, in short, made up of young offenders. It is hard to imagine an idea quite so likely to offend the tabloid press, but it works.

Cahn believes that the over-stretched prosecutor's office makes the youth justice system create long-term criminals, because it tends to ignore the first two cases against teenagers. In fact, they all know they have to be caught three times before anyone takes them seriously. 'Our job was to create a different set of options by creating a new sub-culture – a place where it is safe to say to a peer something as simple as: "Don't do something stupid." We had to teach – or more accurately, we had to enable young people to confront their peers with the obvious: "You knew when you got into that car that there was no way that Mercedes belonged to him."'

Cahn's Youth Court now handles half of all the first time offenders in the city. The juries rise to the responsibility, and use their peer pressure to reinforce sensible behaviour rather than the other way around. The court makes it possible for them to say the things they really think, even occasionally to reach out in a human way to the teenagers who come before them. They have also reduced re-offending for those taking part from 30 per cent to nine per cent. Sticky organizations work because they can turn intractable situations upside down, but – if politicians are the last to understand this – we may keep this idea on the margins for decades to come.

What Cahn's sticky organizations were able to do was to knit together social networks again, person by person, relationship by relationship, and the magic ingredient seems to be that they take an all-embracing view of human skills. They take normal public services and turn them on their heads. The doctors or job centre staff might be concerned about what the person in front of them needs. They usually categorize them according to what they *can't* do. Sticky organizations look at people from the other way round: they ask what they *can* do and then find ways of putting them to work. Something magical happens when human beings interact, using their human skills, when they have never been asked to before.

The point is that people's passivity seems to get in the way of making change happen. Yet organizations tend to be designed to make clients and customers as passive as possible, because they are easier to process that way. Just take the pills, do what you're told, obey the rules, don't shout at the counter staff, tick the boxes and pay at the till. If you are jobless or an asylum seeker, then that passivity is actually enforced. You are not allowed to do anything which might get in the way of finding a job; you must wait for the Kafka-esque processes of the Immigration Agency to take their labyrinthine course – sometimes for years. Strange, and perverse, but true.

But what is the difference between these sticky organizations, and customers co-creating products? What is the difference between Lego and just asking your customers to load up their own shopping trolleys or to park them neatly when they leave the store? The Australian post office recently claimed that getting people to put their own postcodes on envelopes was a radical example of 'co-production' in action. Edgar Cahn provides the clue. For these ideas to make a difference, they have to shift the balance of power a little. Tesco's self-service tills don't have the world-changing ethic that Cahn promotes at the heart of his projects. What Linux and time banks both have in common is that they are designed to turn the world upside down. If it doesn't do that, then getting your customers to help you save money is just manipulative, and very obviously so to them. Blurring the boundaries between inside and outside needs to create at least some equality in the relationship.

That is what managers wedded to the old system don't understand yet. If making organizations more human and effective is just about asking people's opinion, then you don't really change anything. Starbucks and Dell both have websites to get ideas from customers (http://mystarbucksidea.force.com/ideahome and www.ideastorm.com), just as every public service in the UK has to consult more energetically. They need to get clients onto their boards of directors and ask people's opinion even more than they already do. None of this is changing the power structure. People are just being made into more effective supplicants as a result. They are still asking for things which the company or the professionals then have to decide whether to provide. The organization maybe using their customers as resources, but only the brains of their customers – their management advice and nothing more. The power balance is still intact.

There are other ideas which really do seem to be blurring boundaries, sometimes by borrowing community participation techniques and using them inside companies. When IBM organizes its Innovation Jams, getting 100,000 employees and customers to spend a whole weekend on the internet, trying to think through their future products, then – if it is at all successful – something about their relationships will have changed by the end.

That is why, for the old guard, this all looks very untidy and rather risky. It is not an idea that appeals to managers who want to stay in control, or professionals who value their status as slightly apart from their clients. It particularly horrifies anybody who is concerned about the risk that ordinary people pose to each other, given the chance. The American child psychologist Jill Kinney founded the charity Homebuilders to support families who were at risk, so that she could stop children being taken into care. But professionals can't supervise families forever, and it is always a dangerous moment when they let a recovering family fend for itself. So Kinney suggested that neighbours, or people who had lived through similar problems, might take over from professionals at the end and provide more permanent support. Her colleagues were horrified. She was banished by her own charity and forced to set up another one.

The new health, safety and 'safeguarding' regimes make it very hard to involve people with any kind of criminal history, yet they are often the very people you need to warn children about drugs or getting into trouble. If you vet volunteers and weed out anyone with a criminal record, it does get in the way of this kind of public co-creation. 'I've got the sort of criminal record that means the VBS [Vetting and Barring Scheme] will certainly vet and bar me,' wrote Mark Johnson in *The Guardian*. 'Yet it's my criminal record that makes me particularly qualified to work with young offenders. It's my years of drug addiction that give me a special understanding of addicts. It's the changes I've made in my life that offenders and addicts want to hear about. A prison governor told me I can have more effect on his inmates in 30 minutes than he can in three years.'

But probably the last profession to get the idea is going to be politicians. I realized this one sunny morning in 2007, wearing my suit – a rare occurrence – to give evidence on these issues to the Public Administration Select Committee in Parliament. The ideas behind 'co-production' were not completely new, but it seems to be especially difficult for politicians to grasp the idea that people might be more than just passive recipients of their generosity.

I had known Edgar Cahn for a decade or so, and had even watched his projects in action. I had been involved in starting similar ones in the UK. I even wondered whether there might be an explanation here about why the welfare state has been so ineffective over two generations at doing what it was designed to do. The Giant Evils of the Beveridge Report, the founding document of the British welfare state – want, squalor, ignorance, idleness and disease – are still alive and well, after all, and are born again every generation. I wanted to explain that services which people are expected to accept passively don't create social change, because they seem to get in the way of the human connections which make things happen. I wasn't going to ruin things by using the word 'sticky'.

Perhaps the mistake we made, giving evidence to MPs, was straying into their own territory and trying to explain the implications for politics. How do you explain to some politicians that there is more to the job than simply having grateful constituents come to them and asking for things? One MP (no names here) refused to accept this idea completely. 'My people are very grateful,' he told us. We explained how some small parks had been transformed over the past decade by local people forming 'friends' groups, taking some of the responsibility for their upkeep and development. 'More floral clocks,' said another MP. 'That's what we need.'

I looked at him to see how much he was joking, and how much he genuinely misunderstood, and I'm still not absolutely sure. He grinned in a self-satisfied way and leaned back in his chair. It struck me then that it was going to be very hard to persuade him that the shape of public services was upside down. It wasn't about giving things – it was about asking for things back. It wasn't about controlling, or systems or safeguarding. It was about making his local park – as well as his local surgery, police station, housing estate and court – into engines of ferocious regeneration, just by asking people to do things, rather as Debbie Morrison had done at her excitingly upside down school in Stoke-on-Trent.

Find out more

Becky Booth is programmes director of the consultancy Spice (www. justaddspice.org has details of the organization's extraordinary projects).

There is a vast literature about social capital and communitarianism, some of it fascinating (Robert Putnam's *Bowling Alone: The Collapse and Revival of American Community*, Simon & Schuster, New York, 2000), some of it a little depressing (Amitai Etzioni, *The New Golden Rule: Community and Morality in a Democratic Society*, Basic Books, New York, 1997), and some of it worryingly authoritarian (Tony Blair, *The Third Way*, Fabian Society, London, 1998). This has fed into the 'broken Britain' debate in British politics, which – as you will realize from this book – I don't accept. You can read more about Polly Wiessner's work on the anthropology of reciprocity in 'Risk, reciprocity and social influences on !Kung San economics' in *Politics and History in Band Societies* (Eleanor Leacock and Richard Lee (eds), Cambridge University Press, Cambridge, 1982).

I very much also recommend the Demos work on this subject (Sophia Parker and Joe Heapy, *The Journey to the Interface*, Demos, London, 2006). My peculiar session with the Public Administration Select Committee can be found word-for-word at www.parliament.uk under the minutes of evidence for 26 April 2007.

Localize everything

Supplicant (noun): One who humbly entreats.
(*Roget's Thesaurus*, Third Edition, 1995)

Our civilization ... has not yet fully recovered from the shock of its birth – the transition from the tribal or 'closed society', with its submission to magical forces, to the 'open society' which sets free the critical powers of man.
(Karl Popper, *The Open Society and its Enemies*, 1945)

Summary

- We need to devolve decision-making far lower for the institutions that affect us, because then – and only then – can we set free 'the critical powers of man'.
- The enemy is the Supplicant State, where we are constantly begging for things, from public and private institutions, often via call centres that can't help us even if they wanted to.
- What can you do as a government or chief executive wanting to shift the leviathan in a new direction? The answer is that you choose your lieutenants. You have to inspire and cajole. You have to paint a compelling picture of the objective and thrill people about how to make the journey possible.

Imagine yourself in the coffee houses of 18th century Edinburgh, in the elegance of the New Town when it really was new, the civilization of those paved streets, and the intellectual excitement of the Scottish Enlightenment. It was there that the philosopher David Hume first cast doubt on scientific method, peering at ideas about what causes what and finding there was nothing there. All you can do, he said, is say that events tend to happen together. Yet, if we can see nothing causing things under the philosophical microscope, that hands the scientists a big logical problem. It doesn't matter how many times they do an experiment, or watch the sun rising bang on time, it doesn't mean these events are any more likely to happen tomorrow.

Two centuries after Hume was writing in Edinburgh, the Viennese philosopher Karl Popper, a refugee from the Nazis, came up with an interim answer. But, more importantly, he also applied it to politics and organizations. You may not be able to prove what you believe about the world, no matter how often an observation or experiment takes place, but you can *disprove* it. Popper used the example of swans. It doesn't matter how many white swans you see, it still doesn't prove that all swans are white. But if you see a black swan, then you know they are not.

Popper was writing during the Second World War, his home city was in the hands of totalitarians, and he quickly found himself applying this insight to politics too. In doing so, he produced one of the classic 20th century statements of philosophical liberalism, *The Open Society and its Enemies* (published in 1945). He said societies, governments, bureaucracies and companies work best when the beliefs and maxims of those at the top can be challenged and disproved by those below. This has huge implications, not just for effective societies, but for effective organizations too.

Popper was flying in the face of the accepted opinions of the chattering classes at the time. They may not have liked the totalitarian regimes of Hitler or Stalin, but people widely believed the rhetoric that they were somehow more efficient than the corrupt and timid democracies. Popper explained why they were not and why Hitler would lose. Anybody who has read Antony Beevor's classic account of the Battle of Stalingrad, and the hideous slaughter and inefficiencies brought about by two centralized dictators who had to take every decision personally, can see immediately that Popper was right. Real progress required 'setting free the critical powers of man', he said.

The possibility of this challenge – in what he called 'open societies' – is the one guarantee of good and effective government or management. Those human beings at the front line, those most affected by policy, will always know better about their own lives or their own work than those at the top. Open societies can change and develop; closed societies can't. Hierarchical, centralized systems, by their very nature, prevent that critical challenge from below.

This kind of clarity is the antidote to the illusions of control that McKinsey-style measurement brings. It means you can bring to bear more intellectual resources. That is why democracy triumphed over the totalitarians, and why simple, flexible organizations tend to beat hierarchical, over-controlled ones.

We know, thanks to all the rules that went before this one, that this proposition flies in the face of the whole reason for these controls in the first place – that human beings don't know best. Left to themselves, won't they screw up, just as they used to? As we have seen, they will screw up. They will muddle and fuss and sometimes create the most infernal mess – but the likelihood of those mistakes is still better than trying to control their every thought and move. The point about Rule 10 is that there are political implications for all this. It isn't just management control that needs to be light, but political control too. There needs to be leadership, direction, inspiration – from chief executives and from politicians – but their control over the details needs to be light. We need to devolve decision-making far lower for the institutions that affect us, because then – and only then – can we set free 'the critical powers of man'.

But there is rather a difficulty here, at least in the UK. Britain has been one of the most centralized states in the world, despite the Localism Bill 2011, which devolves important powers to local authorities. We are steeped in that a culture dominated by London, Westminster, Whitehall and the City, which means that even those most enthusiastic about localism are stuck with the political language and accepted political solutions that derive from a centralized political culture. It is difficult for British politicians, however much they might want to, to find a new language that can genuinely break free from the old assumptions. 'The housewife of Britain has to accept that the man in Whitehall really does know best,' said Lord Shawcross in the 1940s. 'We are dealing with people who have no initiative or civic pride,' said Newcastle's chief planning officer in 1963, revealing the contempt of the governing classes for the governed. 'The task surely is to break up such groupings, even though people seem to be satisfied with their miserable environment and seem to enjoy an extrovert social life in their own locality.'

Nobody would dare say things like that today, and the second quotation goes some way to explaining how working class neighbourhoods could have been so disastrously redeveloped in the 1960s and 1970s. But the basic idea is deep in the DNA of British government at every level. British government does not believe it has anything to learn from the local, whether it is local people or institutions. There may be a new government in Westminster that uses the rhetoric of localism but, as I write, they are busy instructing local authorities to reduce the number of street signs. They may be right, but it isn't exactly localism.

The reason that Whitehall clutched to itself so many decisions is partly sheer arrogance, but it is also because they believe it is the only way to organize a fair distribution of resources. But even that isn't quite accurate. The NHS is one of the most centralized organizations, and one of the biggest employers in Europe, yet its resources are hideously unequal. You can wander for miles across the East End without any sign of a doctor's surgery. Localized systems can be unequal, but centralized ones can just mean that the elite can organize the resources better for themselves.

That is why centralized decision-making eventually grinds to a halt. Norway celebrated the millennium in 1998 by setting up a commission of five eminent men and women to inquire into the state of their democracy nearly two centuries after independence from Denmark. Their 50-volume report was a frightening wake-up call to the country that their democratic infrastructure was approaching collapse. They found that participation in government and elections was plummeting, that the municipalities – the source of Norway's democratic energy – were being captured by the central state, and power was falling into the hands of unelected technocrats. The result was stagnation.

Norway was relatively late coming to this conclusion. A similar realization in France in 1982 led to power being transferred dramatically from the *préfets* – some of whom are supposed to have broken down in tears at the news – to the mayors of the cities and communes. Other European countries have taken similar dramatic decisions over the past 20 years.

This is not to say that the French system, or any of the other decentralized systems of Western Europe – some of them dating back to the post-war constitutions drawn up by the victorious allies in 1945 – would be precisely right for Britain. The point is that the UK's neighbouring countries and allies realized the damage that the drift towards centralized decision-making was doing, to their own democracy and public services, and reversed the trend.

Britain has been very late to begin following suit, despite the lip service paid to the localism agenda by the governments of Tony Blair and Gordon Brown. David Miliband famously promised schools 'a 25 per cent reduction in inspection ... a 40 per cent cut in data submission', when he was a junior education minister. But when there were crises in child protection, more inspection was always the solution ministers offered – more databases, more registering and more formulaic controls. National politicians need to appear effective. But the local institutions have been corroded; of course they must act centrally, whether it works or not. What else could they do?

In his pamphlet *Big Bang Localism*, the columnist Simon Jenkins used the example of London, which in 1900 elected 12,000 people to run their local services. By the time Tony Blair was elected in 1997, that figure had shrunk to just 2000. The rest had been replaced by 10,000 appointees to boards and quangos, the creatures of Whitehall, running everything from roads and transport to health. The process accelerated after 1997. The extreme disconnection between the decision-makers and the front line – the main factor in the failure of public service reform – was starkly obvious if you compared the structure of what remains of local government in the UK with other European countries.

The average population of the lowest executive tier, usually district or borough councils, is 118,000 in the UK. That compares to 1500 in France, 5000 in Germany and 7000 in the USA. We have 2605 electors on average per elected representative, compared to 116 in France and 250 in Germany. Our local councillors are outnumbered three times over by 60,000 unelected people, appointed by the centre, running 5200 local quangos that still dominate our experience of public services. They were the direct result of the contempt with which Whitehall usually regards its equivalents in local government.

But it is worse than that. More than 100,000 people sit on parish councils in the UK which, unlike most of their equivalents in other countries, have absolutely no power. Elected councillors divide their time as 17 hours with the public per month and eight hours with officials. The quangocracy divides it the other way round: four hours with the public and 11 hours with officials. The first Blair term imposed no fewer than 6000 targets on their poor benighted local councils and quangos, from the number of children adopted to the number of CCTV cameras, many of them conflicting, backed by a battery of cappings, ring-fencing, inspectors and all the panoply of empire. And parallel to the centralization of government power is the centralization of corporate power, the huge semi-monopolies granted to companies such as Tesco and Capita.

'The plenipotentiary state of New Hampshire has only half the population of the capped and cribbed county of Hampshire in England,' wrote Jenkins.

Taken together, this combination of public sector sclerosis and private sector monopoly have been a devastating blow to local life in the UK, helped by punitive health and safety regulations, products of big corporations, central bureaucrats and risk-averse insurers. It has been emptying our institutions of meaning, and rotting away their humanity, corroding imagination, innovation and pride at local level. It is also a far broader problem than the current debate about localism would have us believe. It represents a shift in our status as citizens, a diminution of our individual power, a reduction to the status of suppli-cants to giant and distant organizations, public and private and often a mixture of both. The result is a massive loss of confidence in what local people can achieve and a growing frustration with the slowness of change. This frustration feeds back into more centralization, more scle-rosis, and more centralization again, because it gives the illusion of change without actually achieving anything.

At the heart of all this is a decadent metropolitan snobbery. It is because of the contempt that the City of London feels for industry and small business, and the contempt that Whitehall civil servants feel for their local counterparts. A century ago, this process was predicted by the writer Hilaire Belloc, who called it the 'Servile State'. To be more precise, it is a Supplicant State. We are constantly begging for things, from, public and private institutions, often via call centres that can't help us even if they wanted to.

Things are changing, even if they are changing slowly. A timid version of localism is on the mainstream political agenda at last, and the economic crisis represents a glimmer of hope as well. Highly centralized states come under intense political pressure during recessions as need mounts, and they find themselves even less able to meet it than they were before. A policy promoting local life is absolutely urgent in economic diffi-culties. None of that means it will necessarily happen, but it makes it more likely.

There are also dangers in an approach that is too ideological. Localism will not work if there is nowhere to appeal against local corruption or local idiocy (though the appeals do not have to go through the centre). Nobody should pretend that localism is risk-free. The story of one American town that banned Anne Frank's diary from the local libraries is a warning to anyone pursing this agenda. They decided it was 'a bit of a downer'. Equally, the idea of suddenly devolving power to local councils that have recruited for a generation on the basis of following process and using no imagination is a recipe for more sclerosis.

But the core of the political problem was there when I gave evidence to the Parliamentary Committee (see Rule 9). Politicians like the idea of taking decisions locally because they are politicians. They tend to think that taking decisions is the highest form of human activity. But beyond that, most of the rules above are difficult for them. They tend to involve a loss of control, rather than artificially simplifying complex systems into procedures. Worse, they imply a loss of role – what are politicians for if people are increasingly managing their own organizations? It is frightening for them, if they grasp it, and – all too often – they don't get that far, and stay clinging desperately to the levers of power, determined that they should still affect something.

Devolving power to local people is necessary if we are going to humanize the way we are governed, but it really isn't enough. You could devolve power to factory hospitals or giant faceless schools and be no better off. The whole point about making institutions human-scale is that, when you know the headteacher yourself, you need the imprecision of a league table or target figure that much less. In fact, there are two areas at least where the kind of localism pedalled by most politicians still falls short of what we need. One is economics and the other is about institutions.

There is no point in devolving powers to towns and cities if, in practice, they still have no control over where their spending power flows to, over what businesses set up locally and what effect they have on the local economy. There is no point in localism if towns and cities stay powerless supplicants to the power of semi-monopolies such as supermarkets or giant waste contractors. The massive centralization of decision-making by companies and retailers is an even more powerful force than the centralization of political power. It is also driving the 'ghost town' phenomenon, emptying communities of local post offices, banks and pubs, or transforming the places we live into bland identikit non-places where the economic and buying decisions are taken by distant purchasing executives, and where shopkeeper's jobs have been replaced by low-paid check-out staff.

When economic decisions are centralized you begin to lose the kind of intricate local links between businesses, which keep the money circulating and keep places alive. Multiple retailers can bleed local economies dry of money: earnings do not recirculate unless there is a basic bedrock of locally-owned business to trade with. Local authorities that accept large new retail developments on the grounds that they are worth a certain amount of money forget to ask whether than money is likely to stay put. Then the consumers get less choice than they would with locally-owned shops that are, theoretically at least, able to stock fresh local produce if they want to. Worse, the social glue provided by real local shops that holds communities together is dissipated. When it's gone, we find that it used to prevent crime, spread responsibility, build social capital and even tackle loneliness.

In towns such as Bicester in Oxfordshire, where there are now six Tesco stores, there is very little choice about where you shop or how you shop. But governments have adopted policies that allowed a few retailers to build up semi-monopolies in regional and local markets. The Office of Fair Trading believes any market share above eight per cent to be damaging to the retail supply chain, yet successive governments have allowed Tesco to build up a share of more than 30 per cent.

I don't want to single out Tesco. The centralized banking system is just as bad. The problem with the handful of huge banks that dominate the UK market is not that they *won't* lend locally, but because they *can't* lend using their current infrastructure and systems. They have been consolidated to the point where they point towards the speculative economy and have little local lending infrastructure left. Their lending decisions are taken by computerized systems that, because we are in a recession, naturally recommend against lending. There are no longer bank managers, or local staff with the authority to pick out the success stories, using their knowledge of their local economy.

Our local economies are now in a far weaker position than their American or German competitors, because we don't have the kind of local lending infrastructure that they have. There are only 170 branches per million people in the UK, compared to 520 in Germany and 960 in France. One look at the figures for lending before and after the banking crisis gives the game away. As much as 48 per cent of their business lending was to the property sector, according to Bank of England figures. That was bad enough, but since the crisis, that figure has leapt to 78 per cent (and that is over four years to March 2010). They have no source of local knowledge or expertise about local entrepreneurs or markets. All they can do is lend on the latest property bubble and fuel its trajectory. They understand lending on assets, but can't deal with lending on cashflow. They understand big, but can't grapple with small.

Half the money kept in American banks is in small ones. We don't have that option in the UK, and the combination of this – and the consolidation of retailing, and the official blind eye turned to monopoly power – has been ironing out the humanity of the places we live in. Each closure is bad enough on its own: a quarter of all bank branches and fishmongers' shops disappeared in the 1990s. But when the number of local retailers falls below a critical mass, the quantity of money circulating within the local economy suddenly plummets, as people find there is no point trying to do a full shop in town. We have been losing all the other fine mesh of what makes the places we live human, from open space to community buildings and meeting places. London alone has lost the equivalent of 1428 football pitches, or seven Hyde Parks, since 1989. Wholesalers, the lifeblood of small local shops, closed at a rate of six per week over the past decade. About 36 pubs are now closing every month.

Diverse local economies, where local business can keep money circulating by trading with each other, are more flexible, more able to survive global recession and more innovative, than ones that are dominated by a handful of names. That is why any localism that simply administers government more locally and democratically – but leaves in place the same forces of centralism and giantism in business – leaves people very little better off. They are still supplicants to distant boards of directors just as they were supplicants to distant governments. It isn't human.

There are echoes of this critique from those idealists of the 1970s, such as Herbert Marcuse or Ivan Illich. Or even in the writings of Hilaire Belloc and G. K. Chesterton in the 1920s. All of them warned about the hollowing out of institutions by corporate power and industrial-style systems. But most, if not all of them, were either so melancholic that no solution seemed possible, or so other-worldly that none seemed very likely either. The exception, perhaps, was the Jewish theologian Martin Buber, who derived some of his inspiration from the Israeli kibbutzim in the 1950s and 1960s. It was he who urged us to rebuild the local institutions we need to sustain relationships, especially those economic relationships that make everything else possible – so that we don't have to stay being supplicants to distant systems in quite the same way.

Something is happening, especially now that there are 100,000 on waiting lists for allotments to grow their own food in the UK alone. There are new food co-ops, subscriptions to local farms, community agriculture, community land trusts and local currencies – a whole range of pioneering new institutions which can begin to underpin people's economic lives.

My favourite story, which combines some of these factors, is in an impoverished concrete housing complex outside the Luton, called Marsh Farm. There was little enough happening economically there, except perhaps the welfare cheques. The money came in by cheque and then went straight out again. They got together to work out where their money was going to and found that, between all the households, they were spending £1 million of their money every year on fast food outside the estate. That meant £1 million was leaking out of the local economy.

So they started a new business, employing local people, to provide healthy fast food. Then they leased some unused fields from the council and grew some of the ingredients. And so it goes on. The think-tank where I am a fellow, the New Economics Foundation, is doing similar things with immigrant groups outside Tel Aviv, in shanty towns in

Honduras and Lima, and outside Durban in South Africa, the heart of Zulu territory. We were told by local development agencies in Durban that there was no entrepreneurial flair there. Yet 174 people showed up at our event, and put forward 100 business ideas, 82 of which are now being supported by a business coach there.

Despite what people say, it isn't that there is any shortage of people with ideas. But there *is* a serious shortage of confidence and social networks and institutions, just as Martin Buber suggested. That means not just tens of thousands of little mutuals to run our parks and libraries, but – most of all – economic institutions that can share some of the benefits of the economy much more widely. The American economist and campaigner Gar Alperovitz, who has written widely about Buber's economic ideas, is one of the instigators of what is one of the most exciting local projects anywhere in the world, an innovative combination of local money flows and Spanish-style cooperative businesses. This is the Evergreen Co-operative Development Project, and it is happening in Cleveland, Ohio, which is probably the epicentre of the conflagration that followed the collapse of the sub-prime mortgage market and the banking crisis.

There are two major economic players still active in the city of Cleveland, and they are active because they are in the public sector – the university and the hospital. Both are spending money, but where? And is it making a difference to Cleveland in a way that can improve both their economic prospects? If Cleveland's hospital spends large chunks of its money on contractors elsewhere in the world, or if it banks its money in Wall Street rather than locally, then of course it will mean more local ill-health, greater demand and less local resources to tackle it.

Instead, the Evergreen Project is a way that the hospital can start spending a much larger chunk of its procurement budget in the city. But that much is hardly new. The next part of this project borrows from one of the great success stories of cooperative business, Mondragon in Spain. The Mondragon story dates back to just after the Second World War, when the local Catholic priest, José Maria Arizmendiarrieta, founded the first worker's co-op to employ local people and meet local needs. More than half a century on, there are now 256 linked cooperative businesses, employing nearly 100,000 people and with offshoots worldwide, and they have been doing even better during the global downturn. So much so that the US steelworkers union have signed a long-term agreement to do something similar in North America.

The Evergreen Project aims to do the same in Cleveland, but clustered around and dependent on the hospital, starting with a major sustainable laundry business. The second project is going to be a renewable energy company, again giving local people an ownership stake, and starting with installations on the hospital roof. This is a hugely exciting new departure for localism. It provides a glimmer of an idea about how cities, towns and regions might re-grow their local economies, and make sure that the benefits stay local, and that local people have an ownership stake in them.

It means a little humanizing of the idea of money, regarding it as more like water. It sees local economic recovery not so much in terms of scarce investment, but in terms of making the flow work better. If money is water, then you can direct it, use it for irrigation, build your mills on the streams. You don't just wait around, and wring your hands and say desperately – if only someone would build a river here. Cleveland is using what flows it has for irrigation, and building out from there.

This is a model of the future where human beings are much more able to use what skills they have in the economy, less dependent either on handouts from a distant state bureaucracy or a giant technocratic employer. It is a vision of a kind of economy that doesn't waste people's skills by leaving them idle. It requires action at a national and international level too – we can no longer afford an economy which is designed primarily to cater for corporate *übermensch* in the forlorn hope that somehow their wealth will trickle down to the rest of us. But most of all it requires local institutions, not dependent on distant funders, and as many of them as possible. The age of 'rationalization', of pointless mergers to produce ever bigger corporations, charities or government departments, is not quite over – but the writing is on the wall.

Ivan Illich is a key figure in what is emerging, though a largely forgotten one. He was an Austrian Catholic priest, in the Washington Heights neighbourhood of New York, before he published a series of revolutionary ideas in the 1970s that provided a powerful critique of our institutions, for being so systematized and professional that they undermined our ability to develop ourselves as we saw fit. *Deschooling Society* warned that the education system was making us stupid. *Limits to Medicine* warned that the health system was making us ill.

At the time, his criticisms barely reached the people who run these monster institutions. Now things are different. The debate about spending cuts is obscuring a far bigger political struggle over the future of public services between those who want them to carry on much as they are (except more so) and those who fear that they are failing because they are so systematized. The investment by the Blair and Brown governments may prove to be the last gasp of the old system. Most of the health investment, after all, went into employing more doctors and nurses – when the real threats to our health are elsewhere, and where 80 per cent of NHS spending goes towards maintaining people rather hopelessly in their chronic conditions.

The new revolution is emerging in some of the ideas in this book, but in tiny ways – people learning to take their own blood tests or taking on responsibility for their own disability budgets, as well as the time banks and expert patients that are already beginning to turn the NHS upside down. This is not really about 'choice', which just seemed to embed the bureaucracy ever deeper and ever further out of reach. It isn't actually about decisions, except about your own treatment perhaps. It is about *doing* things (see Rule 8).

So the real localism agenda, beyond the political rhetoric, goes beyond simply decentralizing decision-making, just as tackling the Supplicant State means more than just devolution. It means rebuilding the very engine of public service – the healing and educative relationship between doctor and patient, teacher and pupil. It means rediscovering that ordinary users of services have something to offer that society needs. It means dumping the software the distant institutions use to manage us. And it means rebuilding the possibility of local enterprise and energy to breathe life into our local economies.

So we need to develop a more human kind of localism, capable of explaining why government is so ineffective, why prisons are so useless at reducing crime, why the NHS is so useless at preventing illness, why the welfare state is so useless at reducing poverty and why Westminster is so plodding in its delivery of real change. It means that localism needs to go much further, and include measures to remove central targets and replace them with better recruitment, better training and better local management. It needs to shift central inspection from tick-box computerized checklists to genuine professional mentoring. It needs to re-organize local public service institutions as outreach centres that can manage the vast army of volunteers we desperately need to have any long-term impact on social issues. It also needs to break up and break down the monster institutions, and to break up the biggest companies and re-introduce genuine competition.

This is not so much a paean of praise for people power, so much as a thought about the shape of government we need if we are going to make the human element work again. It means that the objective of politics has to change. It means a shift in political rhetoric, being prepared to devolve responsibility to ordinary people as well as power. It means a radical new offer from politicians to the public. Not 'ask and you shall receive' – nobody believes that any more, least of all the supplicant voters. No more dashing in and sprinkling fairy dust around while everyone cheers. It needs to say: we can achieve these things, but not without your help.

None of this is simple. The model of how things change was obvious when you could imagine organizations like vast machines. Even if it was a delusion, it seemed as if you could just pull a lever and tell someone what to do. What can you do as a government or chief executive wanting to shift the leviathan in a new direction? The answer is that you choose your lieutenants. You inspire and cajole. You paint a compelling picture of the objective and thrill people about how to make the journey possible. You have to set the course. You have to mark the charts and set sail. You have to keep a hand on the tiller and watch the changes in the wind. What you can't do is to row yourself or heave on every rope or take every decision. You have to choose your crew and respect their genius. That is the human element.

Find out more

I have tried to write my own summary of the arguments about a broader kind of devolution of power in *Localism: Unravelling the Supplicant State* (New Economics Foundation, London, 2009). But you can't beat Simon Jenkins' diatribe *Big Bang Localism* (Policy Exchange, London, 2004). The argument about economic localism is partly in *Clone Town Britain* (Andrew Simms, Petra Kjell and Ruth Potts, New Economics Foundation, London, 2006). There is more about this at www.neweconomics.org and www.pluggingtheleaks.org. See also Colin Hines (2002) *Localization: A global manifesto*, Earthscan, London. It is definitely worth reading the American localism gurus, in particular see Michael Shuman (2000), *Going Local*, Routledge, New York.

See also the work of the New Economics Institute (formerly the E. F. Schumacher Society, www.neweconomicsinstitute.org) and the Institute for Local Self-Reliance (www.newrules.org).

The other people I mentioned are Antony Beevor (*Stalingrad*, Penguin Books, London, 1999), Karl Popper (*The Open Society and its Enemies*, Routledge & Kegan Paul, London, 1945), and Martin Buber (*Paths in Utopia*, Syracuse University Press, New York, 1996). See also Gar Alperovitz, 'The reconstruction of community meaning', in *Tikkun*, May/June 1996.

Ivan Illich is worth longer study. See *Deschooling Society* (Marion Boyars, London, 1971), *Tools for Conviviality* (Marion Boyars, London, 1974), and *Limits to Medicine* (Penguin Books, London, 1974).

Finding a new horse

There is nothing more difficult to carry out, nor more doubtful of success, nor more dangerous to handle, than to initiate a new order of things. For the reformer has enemies in all who profit by the old order.
(Niccolo Machiavelli, *The Prince*)

It is bad enough to be shackled by unhelpful practices, but it is even worse to have to pay extra for really expensive shackles.
(Kimball Fisher and Maureen Duncan Fisher, *The Distributed Mind*, 1998)

Summary

- We are reaching a tipping point where the mainstream has become so 'efficient' that almost no business except financial services can survive, and where mainstream corporations tend to get overwhelmed by their own externalities.
- Putting the human element to work will be more expensive initially, but it will cost less in the long-run because it works: effectiveness is what counts, not just narrow efficiency.
- The human element is about unleashing this huge energy of care and attention to detail, not just for the places where we work, but the organizations we work in.

'When you're riding a dead horse,' according to one Native American saying, 'the best strategy is to dismount.' This was how the American prophet of human-scale service, David Osborne, urged people to think differently. 'You don't change riders,' he said. 'You don't re-organize the herd. You don't put together a blue ribbon commission of veterinarians.

You don't spend more money on feed. You get off and find yourself a new horse.'

But what kind of horse? This book suggests that it is a people-powered one, because people make the difference between success and failure in human endeavour. As the People Principle suggests: If you employ imaginative and effective people, especially on the frontline, and give them the freedom to innovate, they will succeed. If you don't, they will fail. Then there is the other question: how do we get to a position where we can have the new horse? This book suggests that, actually, it is on its way anyway.

The stories set out to illustrate some of these rules show that things are changing already. They are the first volleys in a revolution that is going to overwhelm us, and probably – these things being what they are – quite suddenly. It is only a few maverick companies now, but one day we will wake up and find it is mainstream. This Berlin Wall of the mind is beginning to crumble, in business and government and all the weird new spaces in between. Those 20th century giants Ford (pioneers of the assembly line) and General Motors (pioneers of the hierarchical monolith) are teetering on the edge of bankruptcy, pushed aside by companies like Toyota that organize themselves in a more flexible way. Even Toyota has its problems now, brought on by a hardening of the organizational arteries and from clutching decision-making too close. The new generation that will push forward the reforms, to use people's skills and imagination, is beginning to emerge – starting with more human-scale institutions. They are emerging because the old-style monoliths don't work.

When the green economist, and the author of *Small is Beautiful*, Fritz Schumacher, wrote an essay called 'Towards a theory of large-scale organizations', he borrowed a story by Franz Kafka called *The Castle*. Schumacher was drawing on his own experience at the National Coal Board, where they had experimented successfully with devolving autonomy to local subsidiaries. *The Castle* was about the kind of organizations that are never devolved into smaller structures. It describes a land surveyor called Mr K, who has been hired by the authorities, but he isn't sure why or how. His efforts to find out are never very successful. Because of that, he fails to do much in the way of real work. A letter then arrives from the castle. 'The surveying work which you have carried out thus far has my recognition,' it says. 'Do not slacken your efforts.'

Kafka was the master of evoking the nightmare, totalitarian worlds of 20th century management. Distant authorities dispensing praise and blame bearing no relationship to what was actually happening. Deliberate minimization of the human relationships which might make the work bearable. A destructive refusal to welcome the truth and, in the background, a whiff of the gulags. Perhaps we can't afford to recognize how totalitarian so much of our work is, a generation after the fall of the Berlin Wall and the Stasi and their miles of files. When we go through the great portals of a modern corporation, whether it is public or private, through the huge blank marble-floored atrium, past the disapproving eyes of those with the power to let you in the gate, we know deep down that we are entering an almost Soviet world. It is a world of empire and obscure politics, where hierarchy, control and bizarre, distorted information has a huge effect on the lives of the people who work there or depend on it. That kind of hierarchical system eventually collapsed under its own internal contradictions in Eastern Europe.

It is true not just at work, but in the way we are administered too. When the American writer Charlene Spretnak was being driven into Bratislava by two members of the Slovak Green Party on a post-communist visit in 1993, they indicated the dismal high-rise hutches in the suburbs on the way.

'That's socialism,' they said.

Sitting in the back seat, she thought to herself: 'No, that's modernity. Do you think we don't have those sterile, towering boxes in Western Europe, the USA and Japan?'

The point is that you could say the same about most of the cities of the world, whatever their style of government. That is the kind of housing poor people are given. They get rationalized systems in public services too and – if governments believe they can swing it – they will get rationalized, de-humanized healthcare and education too. The problem isn't so much capitalism or socialism, it is that those who rule us are stuck in the cultural assumptions of a previous era: they are modernists long after modernism lost its radical human edge, decades after it was adopted by the international elite, as the style for corporate headquarters and a handful of elite architects whose glass towers shrink the poor humans who live in them or under their shadow.

Those who rule us remain enthralled, as the modernists were, to the ideas of mass-production, and the insights of Taylor and Ford and their 'scientific management'. 'The collapse of the Soviet Union was thus in many ways the collapse of Scientific Management,' said the author, John Raulston Saul. Except that it wasn't. The Soviet Union had obviously failed, but the same family of ideas still dominated Western bureaucracies and corporations too, with their centralized command-and-control hierarchies, their obsession with standardization and their suspicion of the recalcitrant human spirit.

We know that the Soviet system of Brezhnev and Andropov didn't work. We don't always recognize how much the same applies to our own corporate edifices too. We don't quite grasp that, when the Berlin Wall came down in 1989, it still hadn't come down in our own world. The sociologist Richard Sennett started his book *The Craftsman* by describing a visit to the Moscow suburbs in 1988, witnessing the appalling quality of the housing blocks he found there, jerry-built by a demoralized workforce. Even Tom Peters described working in Siemens, which was the inspiration for his first book, as 'the closest thing to working for a communist state'.

We still live with the consequences of some of that demoralization in our own world. We are minor victims of this nearly every day, when we deal with call centres who can't grasp what we need because their software doesn't recognize it. We are victims when we work for these systems, and when the pointlessness of another questionnaire, on another obscure government target or whim, suddenly hits us in the morning as we go to work. But there are bigger victims too, whether we are call centre staff measured for the minutes we take going to the lavatory, or we are 11-year-olds drilled into dullness on summer afternoons to pass the multiple choice questions in our SATs exams.

This is the tragedy of the thing. When we work for the system, it demands that we re-organize our lives and beliefs around an illusion of efficiency. When we deal with it, it demands that we reshape ourselves into the rational one-dimension that is easy to process. That was the insight of the poet David Whyte, who was among the first to make the link between the Soviet system and modern work:

> The old corporate world now passing away had become for us a form of ritual, almost religious life ... It asked us to give up our own desires. To pay no heed to our bodily experience. To think abstractly, to put organizational goals above home and family, and, like many institutional religions, it asked us not to be too troubled by any questionable activity.

The new organizations that are going to take their place – because they have to if we're going to avoid ruining ourselves – are going to need to bring in those aspects of us that are excluded, not just our imagination, but our values and our abilities to forge human relationships. Until they do so, we are going to get angrier.

Polly Wiessner, the co-production anthropologist (see Rule 9), provides a clue about this. Reciprocal human relationships are part of the human condition, she says. We assume they exist. When the organizations we deal with break those ties of reciprocity, betraying their promises, making mistakes and pompously failing to apologize, it enrages us – and for deep, primeval reasons. Why are people so angry these days, a friend of mine who works for a high street chain asked me recently? The answer is that it infuriates us when our reciprocal relationships are betrayed. It happens to us every day and, if human systems are withdrawn, it enrages us even more. There is no excuse for a violent response, even for verbal violence at public service staff who are just as much victims as we are. But those ubiquitous posters, warning us not to insult staff behind the counter – in every railway station and doctor's surgery – are a symptom of this more fundamental human breakdown.

It also creates a kind of vicious circle. The angrier people get at these truncated human interactions – these relationships which pretend to be human but turn out not to be – then the more managers become insecure about their ability to form relationships at all, and the more they retreat into a virtual world of technological control, that actually controls very little.

To put the human element at the heart of organizations, we have to demolish some other assumptions too. One is the idea that we are in a period of unprecedented change. That is what business leaders and politicians tell themselves, partly as an excuse for their creaking organization, partly to explain why they are feeling so exhausted. Of course, we can now pick up the phone to Samarkand or catch a plane and be there by tonight. But none of this has really changed the way we live, certainly compared to the big shifts from an agricultural to an urban economy a century and a half ago.

Compare London to how it was in 1900. The population was much the same, the working hours not that much different. The underground stations and bus routes were called the same as they are now. The buildings are largely the same. The FA Cup final took place in Crystal Palace in those days, but that is hardly earth-shattering stuff. Yet we have been told so many times that change is accelerating that we have come to believe it, and one place where it has slowed down to an absolute snail's pace is the way in which our organizations are structured.

Ricardo Semler, the innovative leader of Semco, describes studying a Manchester linen company which was structured using the familiar hierarchy, with a supervisor in charge of every ten machines, and managers in charge of the supervisors. But, he says, that was in 1633. That basic hierarchy has barely shifted in centuries. Our businesses are still run like that, and our public services are more deferential and hierarchical than ever. It is time we moved on. We have lived with organizations based along these same, military lines for centuries, and we are experiencing a last blast of the old ways – as we embed the tightest bureaucratic controls over people in the software they use at work – before we are forced by sheer intractability to look elsewhere. 'There is something wrong with our ability to change,' says Semler.

What is most peculiar about it is that we have lived through a reforming generation that really believed it was loosening up public services, making them more effective, flexible and entrepreneurial. It was supposed to be a revolution but somehow it turned out the other way round. It was called 'privatization'.

Patrick Daly is an Irish community worker, and an admirer of the philosophers Ivan Illich and Paulo Freire. When he moved to London with his partner in 1995 and failed to find a job, he decided to become a road-sweeper. He liked the idea of working outdoors and enjoyed his job but found himself in the frontline of contract privatization – and it was very different to what he imagined. 'As a privatized road-sweeper, my name is "mate",' he said. 'I don't have an opinion, or if I do have one, I had better keep it to myself; I am not responsible and need continually to be told what to do. I respond only to criticism; praise would be deadly poison. I am lazy and am always ready to doss."

One of his colleagues put it like this: 'They have taken the soul out of my work.' The constant pay reductions, the brutalizing management, he didn't mention – it was the 'soul' he referred to. It wasn't as if being a road-sweeper under local authority management had been exactly nirvana. But what he had then, and his employers had, was continuity, roots into the community and a familiar watching eye on the same streets. They had relationships with their customers, however fleeting and distant. But as private employees under a local authority contract, they became part of a machine that can see no further than the contract, takes no interest in anything beyond the specific deliverables. It may be efficient in a narrow way, but it isn't very effective. How did an idea that was supposed to set free the human spirit end up like that?

The word 'privatization' had a chequered history. It was actually coined as 'reprivatization' by the Nazi Party in the 1930s, as a way of handing over government functions to loyal party officials. The phrase was then borrowed by the great management writer Peter Drucker in 1969, proposing that governments use the talent in other sectors to deliver some of their objectives. 'Government is a poor manager ... It has no choice but to be bureaucratic,' he wrote.

That was the basic idea that was taken up by Conservative thinkers in the 1970s. Sir Keith Joseph's Centre for Policy Studies produced a pamphlet in 1975 which set out the case: 'There is now abundant evidence that state enterprises in the UK have not served well either their customers, or their employees, or the taxpayer, for when the state owns, nobody owns and when nobody owns, nobody cares.' It was a powerful proposition.

In the event, when Margaret Thatcher came to power four years later, she had other things on her mind. There was some tiptoeing towards privatization – the sale of Cable & Wireless and British Aerospace in 1981 – but it wasn't until after the Falklands war and her 1983 election victory that she grasped the sheer power of the privatization idea. It was obvious to anyone who tried to use them that the nation's telephone boxes were largely out of order, and so the privatization of British Telecom (BT) in 1984 was a popular move. As many as 2.3 million people brought shares.

Three years later, the Treasury had earned £24 billion from privatization, and the sale of British Gas provided four per cent of public spending for 1986/1987. The idea of privatizing state industries had spread to France, the USA and Canada. Even Cuba and China were testing it out. The merchant bank Rothschild had set up a special unit to organize privatizations, under the future Conservative frontbencher John Redwood, and Conservative theorists were muttering darkly about selling off the Atomic Energy Authority and the BBC. In fact, selling nuclear power stations was the thin end of the wedge. No amount of spin could disguise the fact that they weren't economic.

The original impetus to sell BT was partly to find private investment for telecoms and partly because of Drucker's original idea that private companies were more efficient. By 1985, that was just one of the benefits – it was also supposed to help employees get a stake in the business, provide wider share ownership and reduce the role of the public sector. All those happened, though one of Redwood's team – another future Conservative star Oliver Letwin – said that actually there was very little evidence for the idea that privatized companies were more efficient. A quick glance at the private health corporations of the USA is enough to cast doubt on this one – their health system costs 13.6 per cent of gross domestic product (GDP), while the public British system costs half that, mainly because a quarter of health spending in the USA goes on the bureaucracy of billing, negotiation and payments. Even so, there was a logic about the idea that added up. Privatizing public services would break those bureaucratic straitjackets, and get a new entrepreneurial energy about the place. They would focus on customers. Things would happen. There would be enterprise and imagination. The human element would weave its magic.

But that didn't happen. The early privatizations led to dramatic increases in effectiveness but, after that, things slowed down. Private corporate giants turned out to be as inflexible and hopelessly unproductive (at least as far as the customers were concerned) as the public corporate giants: they just provided considerably fewer jobs. Often the costs remained much the same. Most privatized services are as sclerotic, inhuman and monstrous as their predecessors were.

The Conservative theorist Ferdinand Mount realized this as early as 1987. 'It is becoming increasingly clear that the regulators have no teeth and the operators no conscience,' he wrote, and so it proved. In fact, the privatized operators were determined to become as much like governments as they could, whether it was the bus operator Stagecoach pulling out of Malawi and Kenya because they couldn't have a monopoly any more, or Railtrack running a unit of 25 staff just to battle with the Rail Regulator.

The first local contracting out on any scale was the rubbish collection in Wandsworth. Within six months, the council was enforcing penalty clauses for poor service, but they gave the same company the contract for cutting the municipal grass because there wasn't anyone else. The same thing happened. Soon the European privatized utilities – E.ON, RWE, EDF, GDF and Tractebel – had become huge institutions, delivering services right across the world. By the 1990s, the American waste company WMX Technologies was running the rubbish collection in Wirral, water in Wessex and the Derby Royal Infirmary. The electricity in Buenos Aires was being delivered by the UK National Grid and its water by Anglian Water and the French company Lyonnaise des Eaux.

A decade later, and the supposedly efficient private utilities are largely in the grip of the same illusions about efficiency as the public sector, with phalanxes of call centres, targets and standards, and are as inflexible as any nationalized industry. 'We are committed to a market economy at the national level, and a non-market, centrally planned, hierarchically managed economy within most corporations,' wrote *Observer* columnist Simon Caulkin.

A letter to *The Evening Standard* in 2007 highlighted the problem. The man who wrote it was describing his girlfriend's flight from Romania to Heathrow by British Airways (BA). BA (a privatized utility) had changed planes at Romania and had failed to put anyone's bags on board. The crew knew this, but still the passengers were allowed to wait hopelessly at the carousel for two hours for their non-existent baggage, which BA staff knew perfectly well was not going to arrive. It actually took three days to get them.

So Peter Drucker was wrong. As it turned out, big companies and big contracts tend to become bureaucratic too. The point wasn't that private was better than public, it was that small was better than big, because small allowed for the human element. Ownership wasn't important, at least in its strict sense. Even so, it was Drucker who provided the clue. Anyone can be an entrepreneur if the organization is structured to encourage them. 'The most entrepreneurial, innovative people behave like the worst time-serving bureaucrat or power-hungry politician six months after they have taken over the management of a public service institution,' he wrote. And so it proved.

The New Rules set out in this book are a description of what comes after privatization, but it has something in common with the original spirit of privatization – it is about releasing human ingenuity and drive. If tackling the dilapidation of our housing stock under the old privatization meant handing public housing over to private contractors, the new privatization means encouraging the power and energy of people, and a combination of creativity, gentrification and DIY – which is actually what has been happening over the past generation. It means unleashing the same force, and same combination, which has revitalized Britain's inner-city primary schools – hundreds of thousands of imaginative and interfering parents taking responsibility.

It means what Karl Popper called 'setting free the critical powers of man'.

A generation after the first privatizations, something similar is stirring. It isn't just the innovations of Timpson, Semco and WL Gore, or the emergence of Wikipedia or Linux, or even the huge innovations in the voluntary sector – social enterprises such as Ealing Community Transport running railways, or Bulky Bob's, which collects old furniture in Liverpool, becoming a major training provider in a new refurbishment sector. It is a sense across every sector that the power of people working together can fulfil people's needs, and create effective but uncategorizable new structures. They are not all about people working face-to-face, or even always about relationships, but they are about setting people free to use their skills as they know best. They are about breaking out of bureaucracy, and the narrowness of 'best practice' processes, just like privatization was supposed to be – but, this time, more fundamentally.

The historian Lewis Mumford believed that human history had been a struggle between two kinds of organization – the 'megamachine' where we are all interchangeable cogs, and the village democracy of cooperation and mutual support. It may be that what we are seeing is a new swing of this long pendulum, but I am inclined to think it is something else: a new

fusion of the two, which is creating organizations that rely on human relationships, but which can network and network again on a global scale.
The reach is potentially worldwide, but the engine is in the repeated
possibility of human relationships and human skills, working together, in
network linked to network, and relationship linked to relationship.

The innovator Robin Murray calls this the 'new social economy'. It
includes the whole gamut of new kinds of organizations from social
enterprises to time banks, from Linux to eBay, but it also includes the
revolutionary new structures inside companies. It is emerging without
any kind of guiding hand, without much help from the state and with
almost no help from mainstream business advocates, though – as we
have seen – there are businesses involved in it. The huge failures of the
mainstream system – the struggling airlines and manufacturers, and the
staggering failure of the banks – are opening up a chink to allow these
innovators in.

Murray described this as an economy, but very unlike the old model
based on production and consumption. Instead, the 'new social economy'
uses 'distributed networks to sustain and manage relationships', blurs
the boundary between producers and consumers, emphasizes repeated
interactions such as care and maintenance, and derives from a strong
sense of values. He argued that the conditions are beginning to emerge
which are likely to accelerate this social economy, because of combination of a business crisis (a dysfunctional financial system) and a social
one (obesity, diabetes, an ageing population and much more besides).

The old ways of doing things, which minimize human interaction,
continue to struggle on, and what will replace it has not arrived at any
scale. As the novelist William Gibson said, the future is here already, it is
just unevenly distributed. What the new social economy, and the human
dimension, have going for them is that they provide a potential answer to
some of the ruinously intractable problems that the existing system has
to face. They do so at a time when there are precious few of these
answers available.

Murray wrote about the 'extraordinary spirit of innovation' that is
emerging behind this new economy, of the 'sense of a pressure cooker',
as the new ways forward are held back by sclerotic systems and scarce
resources. What is going to make the difference is that we are reaching a
tipping point where the mainstream has become so 'efficient' that almost
no business except financial services can survive and where mainstream
corporations get overwhelmed by their own externalities. Then the mainstream will shift very quickly to find new ways of putting people back at
the heart, so that we can all survive.

We are about to live through a generation which will genuinely put Michael Hammer's maxim into effect. 'Don't automate, obliterate,' he said, and that is what is going to happen to our institutions as they exist now. They will be forced to tear down their bureaucracies, and their artificial divisions between inside and outside, and turn themselves inside out. That means a generation of urgent innovation, to find new organizations that fit into no accepted categories, but which work – so that they provide us with the food, health, equipment, energy and friendship we need, and the wealth too, while the mainstream (what remains of it) grinds slowly to a halt. It isn't exactly privatization – public and private sectors have now merged beyond unravelling – but it is the radical devolution of work to people, aware that their innate skills are their main tool to make things happen.

It isn't quite enough just to will this to happen. We have to be a little more heretical, especially at work. In fact, when you think of all those disasters and failures that happened because people weren't heretical enough – from Iraq and Vietnam to the demise of the music industry – it is our positive duty to be. The maverick business writer Seth Godin describes people who fail to think for themselves, and be the heroes of their own lives, as *sheepwalkers*. 'I define sheepwalking as the outcome of hiring people who have been raised to be obedient and giving them brain-dead jobs and enough fear to keep them in line:

> We've mechanized what we could mechanize. What's left is to cost-reduce the manual labour that must be done by a human. So we write manuals and race to the bottom in our search for the cheapest possible labour. And it's not surprising that when we go to hire that labour, we search for people who have already been trained to be sheeplike.'

The real tragedy here is that the process of transmogrification into sheep begins in schools. These New Rules are never going to work quite as they should until we have an education system that genuinely trains people to make things happen for themselves. We need this kind of training for a new kind of leadership at every level, for inspiration and human connection, rather than the slavish manipulation of processes. It also requires schools that can equip people with the ability to connect, imagine, create and innovate, rather than the basic wage-earning skills they are taught at the moment. Schools and colleges must be able to foster inspiration, and no amount of sub-dividing curriculums into learning intentions will do this.

This is going to be expensive, and this is the remaining problem. How can we pay for shifting to the human element, either in companies or in government? How can we afford to pay for more people at the frontline, and to give them the whole new training that is required?

The evidence of the three companies at the heart of this book, Timpson, WL Gore and Semco, is that the People Principle can be an important ingredient for success. Ricardo Semler's evidence is that self-managing teams can also cut direct costs. Semco managers were astonished when they first experimentally split one of their big plants into three. They had expected to need more staff – each unit would need their own service staff, when they had shared them before. But, once the three new units had chosen the staff they wanted, the head count was actually less than it had been before. The evidence that small units are more productive (see Rule 4) confirms this.

But the real problem is likely to be in the public sector, where managers are most confused about the difference between efficiency and effectiveness. Once you start looking more closely at the costs to public services of putting process before people – either their staff or their customers – it is clear that taking people out isn't financially very efficient either. The systems thinker John Seddon (see Rule 6) estimates that the split between front office call centres and back office experts is inflating costs by about 20–40 per cent in planning and road repairs, 20–30 per cent in the administration of housing benefits and 30–40 per cent in care services – and something similar in businesses using the same systems. These are unproven, but are borne out by some of the work of his Vanguard consultancy.

There are other costs we can set against the investment in people we need. There are the cost of mistakes from inflexible systems and the cost of doing the wrong things. Up to 5000 people a year die from infections caught in hospitals, and they are now affecting 100,000 people a year. One person in ten who is admitted to a UK hospital now ends up suffering 'measurable harm', whether it is from mistakes, bugs, faulty equipment or drug side-effects. Longer hospital stays as a result cost £2 billion to £3 billion a year. These can't be avoided completely, but the evidence is (see Rule 3) that systems which try to control staff contribute to their mistakes.

It is impossible to work out what these costs might be in any given organization. What proportion of the public sector heating or lighting bill is unnecessary because nobody feels responsible or nobody knows the local situation? What are the costs of those mistakes, that we've all encountered, when local authorities paint the windows of a council estate just two weeks before the windows were due to be replaced? What are the costs of misunderstanding when people are taken out of systems, compared to the costs of misunderstanding when people are left in? We don't know, though the circumstantial evidence is that the mistakes are bigger when common sense is removed.

Then there are the rising costs when nobody trusts each other, the costs of failing to engage in a human way when things go wrong (see Rule 3). A 2002 study of car-makers by Brigham Young University in the USA and the College of Business Administration in Seoul looked at 350 relationships between buyers and suppliers. They found that it cost the least trusted buyers six times as much as the most trusted in extra negotiation and compliance costs.

There are also the rising costs of demoralization. Imagine you are a social worker responsible for child protection, and you had to spend 80 per cent of your time operating the Integrated Children's System (ICS) database, which controls professionals and encourages them to refer cases, leading to huge numbers of referrals, and enormous administrative burdens – when you originally became a social worker because you wanted to use your human skills. I met a former NHS manager recently, who had just been on the senior leadership course for future chief executives, and it was enough to convince him he must work somewhere else. 'The current performance management culture has a certain idiocy about it that in the end drives out intelligent people,' he said.

There are other huge costs which are passed on to other organizations when government services are so efficient that they don't do the job properly. One advice centre found that 94 per cent of its cases were about mix-ups caused by the Department for Work and Pensions. That is where the government grants for advice go. And bigger still, there are the costs of corroding the networks of family and neighbours that underpin everything else, and the failure to tackle causes. All of these cost us vast sums of money, because our organizations are too timid, too inward-looking and too hidebound to change. These all apply to public services, but they apply just as much to business, and we pay one way or the other.

'You put your hands under other dryers, rub them a bit, then give up and wipe your hands on your trousers,' said the British inventor James Dyson. 'It's something that's always annoyed me.' Dyson is best known for his successful and rather expensive vacuum cleaner. It is successful because, unlike so many other vacuum cleaners on the market, it works rather well. The same goes for his Airblade dryer, an alternative to handdryers, which began appearing in public lavatories in 2007.

The old hand-dryers used to take up to 44 seconds to work – or so Dyson claims. Actually, I have encountered many, especially on trains, that would happily maintain the dampness on your hands at great expense for hours under a pathetic breath of tepid air. The Airblade works differently: it scrapes the water off your hands with a powerful jet of cold air, and claims to dry them in ten seconds. But, and here is its relevance to the human element: it uses 80 per cent less energy to do it. It uses more air but in an effective way. It *works*, so it costs less to run.

This is the approach we need to use if we are going to put the human element into practice more effectively. The Airblade approach means that, even though we will need to employ more professionals and train them completely differently, and we will need to sub-divide our businesses, hospitals, courts and schools, it will be more powerful and effective to do so, and that means lower long-term costs. Smaller schools and a huge increase in volunteering are all going to cost money. It is going to cost more to nurture relationships between teachers and pupils, doctors and patients, and it is certainly going to cost a great deal to train and recruit, and then create, a whole generation of super-catalysts. There is no doubt that a health service based on the human element will cost more than one that isn't. It will cost more in the short-term, but the Airblade approach is an answer to those who can no longer believe in the possibility of change any more. It will cost less in the long-run because it works.

We have become so used to the sheer intractability of the problems we face. We watch an endless succession of innovations marched out to tackle them, many of them basically technological which miss the point entirely. We watch while these innovations struggle along until pushed aside by the next one, but have little if any impact beyond the few individuals affected. We have come to the point where we don't expect change, and are suspicious of it if it is offered us, as if those who suggest it can't quite understand the full complexity of the problems we face. The issue of change, and whether it is possible – personally or politically – is our own generation's intractable issue. We all struggle with it.

The human element offers a way forward. It is expensive, but it is affordable – not just because it will reduce costs over time (though it will), not just because the current system is hugely wasteful (though it is), but because without it the costs of failure and sclerosis threaten to overwhelm us. I am all too aware that this isn't a complete answer: I can't estimate the savings we will make from putting the human element wholeheartedly into practice. All I can say is that, in business and in public services, they are huge.

The human element is no guarantee of success. There will be local abuses, corruption, even occasionally bullying and racism. There will be inequalities and bad management and all the rest. It doesn't remove the need for management or oversight or inspection, though it does imply that this also needs to be based on ongoing relationships. But we have those problems already, without having the benefits of the human element making things work. We are coming to the end of a period when organizations were streamlined, as if they were manufacturing assembly lines, using the rhetoric of 'modernization' or 'rationalization', which has ironed out the very wrinkles that made things work effectively.

But the human element makes things possible again. Not by the dramatic moving of resources in the old industrial fashion, but by creating spaces where transformative relationships can emerge. That may seem small-scale and inadequate to the challenge, but the whole point about the People Principle is that small + small + small = big. It opens up the possibility of a combination of caring and attention to detail which can transform situations and may actually be the only thing that can.

There is something almost spiritual about this possibility. Susan Witt is a former literature teacher who is a key figure in the American organization dedicated to building a more human economy, the New Economics Institute. She describes an almost mystical experience when she was driving to see the poet Wendell Berry, who still works the farm in Kentucky that was run by his grandfather. His poems are testament to the enormous mental energy he puts into his own land.

Susan missed the entrance, near the Kentucky River, and suddenly found herself in what she called a 'shiny landscape':

> You know how, as a traveller, you sometimes see things in a way that you can't see when you've been there every day? You have a fleeting vision of a place more intimately, more deeply. And there I was in this shining landscape. The shimmering leaves, the water glistening, and it wasn't just the sunshine. There was a different character in that landscape, and I thought, 'What's happening?' Of course, we had reached Wendell's farm and I hadn't recognized it. I missed the drive and had to turn back.

It was a deep and intimate vision of the land. She thought about the experience for some time, and came to the conclusion that 'it was a landscape well-observed, but not just observed in a distant way like a reporter might describe. It was an observation that was penetrated with love for the place, a love that could then lead to action'. Something about Berry's day-to-day care for the details of his land had transformed it and somehow made it more alive.

'It was Martin Buber who said that our responsibility on Earth is to waken the sleeping spirit and raise it from stone to plant, from plant to animal, from animal to speaking being,' said Susan. 'Wendell's shiny landscape helped me understand what Buber was describing.'

This very individual experience seems to be to carry a great truth, not just about the landscape, but about the places we work more generally. The human element is about unleashing this huge energy of care and attention to detail, not just for the places that we work, but the organizations we work in – a small echo of those cared for flowerbeds in railway stations long ago, but much more than that. It beckons us towards a vision of enlivened organizations, which are small and human enough to generate this energy, and broad enough – each one with hundreds of intricate, involved and committed relationships dedicated to the work – to unleash a similar shimmering in what we do, to waken the sleeping spirit in our empty and echoing institutions, and through each relationship, and on and on, to create a live, shimmering world.

Find out more

David Osborne's rhetoric about changing horses is from his book with Peter Hutchinson (*The Price of Government: Getting the Results we Need in an Age of Permanent Fiscal Crisis*, Basic Books, New York, 2004). Anyone who wants to know about people's impulse towards pride in their work should read Richard Sennett's important and civilizing book, *The Craftsman* (Allen Lane, London, 2008). It is also worth looking to see how far official thinking has come. See, for example, the Cabinet Office report *Power in People's Hands: Learning from the World's Best Public Services* (Cabinet Office, London, 2009), some of which reflects some of this thinking – some of which definitely doesn't.

The Kafka story referred to here is *The Castle* (Schocken, New York, 1998). E. F. Schumacher's essay was published as 'Towards a theory of large-scale organization' (*Management Decision*, vol 25, 1993). Charlene Spretnak's visit to Bratislava is in her book *The Resurgence of the Real* (Routledge, London, 1999), I have written some of these ideas about modernism in more detail in an essay in *Town & Country Planning* magazine ('Everything today is thoroughly modern – or is it?', May 2006). For more about making relationships more central at work see *The Relational Manager* (Michael Schluter and David John Lee, Lion Hudson, Oxford, 2009).

David Whyte's wonderful book is called *The Heart Aroused: Poetry and the Preservation of the Soul in Corporate America* (Doubleday, New

York, 1994). Patrick Daly's article on being a privatized road-sweeper first appeared in *The Tablet* (3 May 1997).

There is increasing literature about the costs of distrust, but the two examples here were taken from Dov Seidman's book *How* (see Rule 2). The study of system thinking in the advice services is *It's the System Stupid: Radically Re-thinking Advice* (AdviceUK, London, 2008). The core of the argument is about the non-existence of economies of scale, for which see John Seddon's article on 'Why do we believe in economies of scale?' at www.systemsthinking.co.uk.

Finally, I quoted from two of the great prophets of our day. They are Robin Murray (*Danger and Opportunity: Crisis and the New Social Economy*, NESTA, London, 2009) and Seth Godin, whose 'sheep-walking' concept is on his blog at www.sethgodin.com. Susan Witt is at www.neweconomicsinstitute.com. Wendell Berry's lecture 'People, land and community' was published by the E. F. Schumacher Society, Great Barrington, 1981.

Index